Evaluation of Alternative Price and Credit Policies for Spare Parts

Addressing Price Changes and Reliance on Credits

ELLEN M. PINT, ADAM C. RESNICK, MARYGAIL K. BRAUNER,
AISHA NAJERA CHESLER, KENNETH J. GIRARDINI, ERIN N. LEIDY,
CANDICE MILLER

Prepared for the United States Army
Approved for public release; distribution unlimited.

RAND ARROYO CENTER

For more information on this publication, visit **www.rand.org/t/RRA2448-1**.

About RAND

The RAND Corporation is a research organization that develops solutions to public policy challenges to help make communities throughout the world safer and more secure, healthier and more prosperous. RAND is nonprofit, nonpartisan, and committed to the public interest. To learn more about RAND, visit www.rand.org.

Research Integrity

Our mission to help improve policy and decisionmaking through research and analysis is enabled through our core values of quality and objectivity and our unwavering commitment to the highest level of integrity and ethical behavior. To help ensure our research and analysis are rigorous, objective, and nonpartisan, we subject our research publications to a robust and exacting quality-assurance process; avoid both the appearance and reality of financial and other conflicts of interest through staff training, project screening, and a policy of mandatory disclosure; and pursue transparency in our research engagements through our commitment to the open publication of our research findings and recommendations, disclosure of the source of funding of published research, and policies to ensure intellectual independence. For more information, visit www.rand.org/about/research-integrity.

RAND's publications do not necessarily reflect the opinions of its research clients and sponsors.

About This Report

This report documents research and analysis conducted as part of a project entitled *Setting Standard (Non-Creditable) and Born-On Pricing for Reparables*, sponsored by U.S. Army Forces Command. The purpose of the project was to (1) determine the potential impact of eliminating reparable credits and the institution of a single price for reparables that would remain in place from the moment the requiring unit obligates funds and (2) outline a new model for reparable pricing that would provide units with predictability in budget execution, make transactions become a normal supply turn-in transaction without concerns over credits, enable the supply chain, and foster commanders, leaders, and soldiers to focus on preparing for combat versus managing credits and changing prices throughout the year.

This research was conducted within RAND Arroyo Center's Forces and Logistics Program. RAND Arroyo Center, part of the RAND Corporation, is a federally funded research and development center (FFRDC) sponsored by the United States Army.

RAND operates under a "Federal-Wide Assurance" (FWA00003425) and complies with the *Code of Federal Regulations for the Protection of Human Subjects Under United States Law* (45 CFR 46), also known as "the Common Rule," as well as with the implementation guidance set forth in DoD Instruction 3216.02. As applicable, this compliance includes reviews and approvals by RAND's Institutional Review Board (the Human Subjects Protection Committee) and by the U.S. Army. The views of sources utilized in this report are solely their own and do not represent the official policy or position of DoD or the U.S. Government.

Acknowledgments

We would like to thank the sponsors of this research, LTG Leopoldo Quintas, Jr., and LTG Paul Calvert, for their guidance on this project. We also thank our action officer, Brian Haebig, and his colleagues in the U.S. Army Forces Command G-4, G-8, and G-3/5/7 for providing feedback on this research and helping us schedule interviews with other Army stakeholders. In addition, we thank the Army stakeholders and other personnel from the Office of the Secretary of Defense and the Department of the Air Force who participated in interviews and provided their valuable perspectives. We also thank our RAND colleague Thomas Light and Edward Keating of the Congressional Budget Office for their thoughtful reviews that helped improve the quality of this document.

Summary

Army units receive Operations and Maintenance, Army (OMA), funding to cover the costs of spare parts and other resources required for training. They pay the standard price (latest acquisition cost plus a surcharge) to purchase depot-level reparable parts (DLRs) from the Army Working Capital Fund (AWCF) and receive a credit (equal to the standard price minus the repair cost) for returning unserviceable DLR carcasses to the wholesale supply system for repair. During fiscal years (FYs) 2020–2021, the AWCF experienced cash flow problems because of a decline in sales caused by the coronavirus disease 2019 (COVID-19) pandemic, which occurred at the same time the AWCF had expanded purchases of spare parts to address supply availability problems. Among other actions taken to restore the cash balance, the AWCF turned off credits for most DLRs from March through September, 2021. However, Army units continued to return DLR carcasses to the supply system, even though they no longer had a financial incentive to do so, in response to return transactions created by the Army's retail supply system, Global Combat Support System-Army (GCSS-A).

This experience raised the questions of whether credits are needed to incentivize carcass returns and whether other changes in price and credit policies could reduce the financial management workload in Army units. For example, if the price of a DLR or consumable spare part changes between the time the part is ordered and when it is issued to the unit, the unit is charged the price at the time of issue. At the end of each FY, all outstanding requisitions are revalued at the new prices and credits, creating unmatched transactions (UMTs).[1] If prices go up, units must have sufficient prior-year OMA funding to pay the higher prices. In addition, units must monitor whether each carcass return is matched to an issue of the same item to receive credit.

Therefore, U.S. Army Forces Command (FORSCOM) asked the RAND Arroyo Center to evaluate alternative price and credit policies, including single price (SP), which would eliminate credits and set a price equivalent to the standard price minus the credit, and born-on pricing (BOP), which would set the price at the time the unit orders the part and obligates the funding. In coordination with the sponsor, we developed additional price and credit policies that would help address financial management workload associated with price changes and reliance on credits, and a set of evaluation criteria to compare the alternative policies. Our research approach included a review of policies and regulations associated with price and credit policies in the Army and the other services, guided discussions with Army stakeholders and other subject-matter experts, and analysis of GCSS-A data to estimate the effects of policy changes on unit budgets and financial management workload.

---

[1] A UMT occurs when the amount of funds obligated to purchase an item from the supply system is different from the amount of funds disbursed to pay for the part, i.e., when the price changes between the time the customer ordered the item and the time it was received.

Courses of Action and Evaluation Criteria

We evaluated eight alternative price and credit policies, defined as follows:

- **BOP A**: A U.S. Department of Defense (DoD)–wide policy that would establish the price when the part is ordered and would apply to both Army-managed items (AMI) and non–Army-managed items (NAMI).
- **BOP B**: Price changes would be absorbed in the AWCF for AMI and NAMI transactions that pass through the AWCF.
- **BOP C**: Price changes would be absorbed in the AWCF for AMI only.
- **Raising the threshold for automatic processing of UMTs (RT)**: FORSCOM would automatically process UMTs with price changes up to $750 to reduce financial management workload; however, prior-year funding would still need to be available in the correct line of accounting.
- **SP**: Sets a price for DLRs equivalent to the standard price minus the credit and relies on supply system discipline for carcass returns.
- **Exchange pricing (EP)**: Sets a price for DLRs equivalent to the standard price minus the credit but charges the unit a delta bill equal to the credit if it fails to return a carcass within a fixed period of time, such as 60 days.
- **Moving the point of sale to the job order (MPS)**: Incorporates shop stock into the AWCF and creates one-for-one issues and returns of DLRs to reduce problems associated with unmatched returns.
- **Free issue (FI) of DLRs**: Units would no longer receive budgets for DLRs, and funding would go directly to U.S. Army Materiel Command (AMC) to purchase and repair DLRs.

We used the following evaluation criteria to compare the alternative courses of action:

- **Effects on unit buying power**: Does the course of action reduce financial uncertainty associated with price changes or improve the steady execution of unit budgets by reducing reliance on credits?
- **Effects on financial management workload**: Does the course of action reduce workload associated with UMTs or tracking credits and unmatched returns?
- **Effects on DLR demands and carcass returns**: Would the course of action increase demand for DLRs and/or reduce the number of carcass returns?
- **Effects on financial stability of the AWCF**: Does the course of action increase risk to the solvency of the AWCF or the functioning of the wholesale supply system?
- **Required approvals by other Army commands, other services, DoD, or Congress**: Are there any likely objections to the course of action by approval authorities?
- **Required changes to legacy or future enterprise resource planning (ERP) systems**: How difficult or costly would it be to implement changes required to implement the course of action?

- **Required changes to budgeting or funds distribution**: Would the Army need to make changes to the budgeting and distribution of funding for DLRs?
- **Effects on other services and Army reserve components**: How would the course of action affect the other services, the Army Reserve, and the Army National Guard?

These criteria could be weighted to emphasize particular concerns, such as potential effects on unit readiness or cost to the Army, but, in practice, the results of the analysis seemed clear without any weighting of the criteria.

Evaluation of Courses of Action

Our evaluation of the eight courses of action is summarized in Table S.1. Green plus signs indicate positive effects, tildes indicate minimal or no effect, and red *x*s indicate potential problems or drawbacks. Among the alternatives focused on reducing the effects of price changes (BOP A, BOP B, BOP C, and RT), BOP A, DoD-wide BOP, would be the most effective, but it was rejected by the other services and the Defense Logistics Agency in 2019, so it has little chance of being implemented in the near future. For FORSCOM, we estimated that BOP B would cover 95 percent of UMTs and 98 percent of the dollar value of price changes and would require an increase in the AWCF surcharge of 0.72 percent. However, it is less effective for

TABLE S.1
Summary Assessment of Courses of Action

Evaluation Criteria	BOP A	BOP B	BOP C	RT	SP	EP	MPS	FI
Unit buying power	++	++	+	~	++	++	~	N/A
Financial management workload	++	++	+	+	++	+	+	+
DLR demands and returns	~	~	~	~	~	~	~	XX
Financial stability of AWCF	~	~	~	~	?	~	~	N/A
Required approvals	XX	+	+	++	X	+	+	XX
Changes to ERPs	X	X	X	+	+	X	X	??
Changes to budgeting, funds distribution, and/or TRM	~	~	~	~	~	~	~	XX
Other services and Army RC	+	~/+	~/+	~	~/++	~/+	~/+	~/XX

NOTE: Green plus signs indicate that a course of action would have positive effects on the evaluation criteria, tildes indicate minimal or no effect, and red *x*s indicate potential problems or drawbacks. A question mark means the effect is uncertain. Doubled symbols indicate a stronger effect. In the bottom row, when effects are different on other services and the Army RC, the first symbol indicates the effect on other services and the second indicates the effect on the Army RC. For example, BOP B would be neutral for the other services and have a benefit for the Army RC. N/A = not applicable; RC = reserve component; TRM = Training Resource Model.

commands with a smaller share of transactions going through the AWCF, such as the Army reserve components. BOP C would cover 15 percent of FORSCOM UMTs and 73 percent of their dollar value, with similar benefits for other Army commands. RT would cover 90 percent of FORSCOM UMTs, but financial managers would still need to ensure that sufficient prior-year funding was available in the correct line of accounting. However, RT is the easiest to implement in existing ERP systems, and the decision would be internal to FORSCOM.

The remaining courses of action (SP, EP, MPS, and FI) are intended to address workload related to monitoring credits and unmatched returns. Our analysis of GCSS-A data on returns of DLR carcasses during the credit shutoff did not find any significant effects of removing the financial incentive for returns. This result suggests that SP could be a feasible alternative, and it is also likely the easiest course of action to implement in existing ERP systems. However, if AMC and other approval authorities have concerns about removing the financial incentives for carcass returns, EP would function in a very similar way, although it would be more complex to implement in logistics and financial management ERP systems. MPS would improve the visibility of shop stock to the wholesale supply system and reduce unmatched returns but would not reduce reliance on credits. FI is potentially risky because units would no longer have budget constraints on ordering DLRs. In addition, changes to budgeting, funds distribution, and ERP systems would be needed if funding for DLRs went to AMC instead of units and financial transactions were not required for AMI but were still needed for NAMI.

Recommendations

To address price changes, we recommend that FORSCOM raise the threshold for automatic processing of UMTs to price changes up to $750 in the short term, because it is relatively easy to implement in existing ERP systems. Other Army commands could also implement this course of action if they think it would be beneficial. In the longer term, we recommend that the Army implement BOP B, absorbing price changes for AMI and NAMI transactions that pass through the AWCF. This policy would eliminate almost all UMTs for FORSCOM and would require a relatively small increase in the AWCF surcharge. However, the policy would be complex to implement in existing ERP systems and would be less beneficial for the Army reserve components.

To address reliance on credits, the most promising courses of action are SP or EP. SP could most likely be implemented in existing ERP systems by changing the catalog price paid by Army customers. However, there might still be concerns that returning carcasses could become a lower priority for units, and there would be fewer carcasses available for depot-level repair programs. Therefore, the Army would need to continue to monitor carcass returns to ensure that they remain in balance with DLR issues. EP would preserve the financial incentive to return DLR carcasses, but it would be more difficult to implement in ERP systems than SP.

Contents

Figures and Tables

Figures

Tables

CHAPTER 1

Introduction

The cost of spare parts is a significant portion of the Operations and Maintenance, Army (OMA), funds allocated to units to complete training. Depot-level reparable parts (DLRs) dominate parts costs and come with the additional requirement of ensuring timely retrograde of carcasses (i.e., the unserviceable parts that were replaced) for repair and to receive credit. Price and credit policies in the Army have shifted over time with the implementation of new information systems and must meet the additional goals of maintaining the financial health of the Army Working Capital Fund (AWCF) and remaining in adherence with Office of the Secretary of Defense (OSD) financial management policy and regulations. Currently, Army customers buy DLRs from the AWCF at the standard price (latest acquisition cost of the item plus a surcharge to cover the operating expenses of the AWCF) and receive a credit (equal to the standard price minus the repair cost) when they return a carcass.

During fiscal years (FYs) 2020–2021, the AWCF experienced cash flow problems because of a confluence of events, including an effort to expand inventories to improve supply availability, followed by a reduction in training activities caused by the coronavirus disease 2019 (COVID-19) pandemic. The resulting reduction in sales of DLRs reduced the amount of cash available to pay the AWCF's contractual obligations. To restore the AWCF's cash balance, the AWCF stopped issuing credits for the majority of DLRs from March to September 2021. This policy reduced units' buying power because credits replenish their budgeted funds and allow them to purchase additional DLRs and consumable repair parts. However, for the most part, units continued to return carcasses without credit, because the Army's retail supply system, Global Combat Support System-Army (GCSS-A), continued to generate return transactions when units ordered DLRs.

This experience raised the questions of whether credits are needed to incentivize carcass returns and whether other changes in price and credit policies could reduce the financial management workload in Army units. U.S. Army Forces Command (FORSCOM) has several concerns with current price and credit policies. First, unit personnel spend a considerable amount of time managing funds and dealing with the uncertainties associated with changes in prices and credits both within and across FYs. Second, units must monitor whether they have received credits for carcasses and whether each carcass return is matched to an issue of the same item. Third, if prices and credits change during the year of execution, there can be uneven impacts across units as operations tempo (OPTEMPO) varies, because budgets are set in advance based on expected parts demands and price and credit values.

Therefore, FORSCOM asked the RAND Arroyo Center to examine the potential effects of eliminating credits for DLRs and instead charging a single price (equivalent to the standard price minus the credit) that would remain in place from the time the unit orders the parts and obligates the funds.[1] The research team also analyzed other alternative price and credit policies for DLRs and consumables to identify options that would provide units with greater predictability in budget execution, simplify turn-in transactions to reduce reliance on credits, help the supply chain operate more efficiently, and free up the time commanders, leaders, and soldiers currently spend on managing credits and changing prices throughout the year.

Research Approach

We used a mixed-methods approach to evaluate alternative price and credit policies, including a document review, guided discussions with subject-matter experts, and data analysis. We reviewed selected references on price and credit policies in the Army and the other military services, as well as analogous systems used in the private sector to allocate resources within organizations with decentralized decisionmaking. We also examined portions of the DoD FMR pertaining to working capital funds and other DoD instructions and directives related to materiel management. In addition, we reviewed the budget submissions of the Army, Air Force, and Navy Working Capital Funds (WCFs) and other documents referring to DLR management and the price and credit policies of the individual services.

We held discussions with subject-matter experts in FORSCOM G-3/5/7 (Plans, Operations, Training), G-4 (Logistics), and G-8 (Programs); the Office of the Deputy Chief of Staff, G-4; U.S. Army Materiel Command (AMC) G-3 (Operations); the Offices of the Deputy Assistant Secretary of the Army for Cost and Economics and Army Budget; the Offices of the Under Secretary of Defense, Comptroller (OUSD[C]) and the Deputy Assistant Secretary of Defense for Logistics (DASD[L]); Enterprise Business Systems—Convergence (EBS-C); III Corps G-8; and the Office of the Deputy Assistant Secretary of the Air Force for Budget (SAF/FMB). These discussions helped us understand the perspectives of different stakeholders that are affected by Army price and credit policies and the organizations that set financial management policies, the financial management workload created by current policies, changes to logistics and financial management information systems that might be required to implement price and credit policy changes, and the Air Force Working Capital Fund's (AFWCF's) pricing system.

---

[1] The U.S. Department of Defense (DoD) Financial Management Regulation (FMR) currently specifies that supply activities can charge the price that is in effect when the item is issued from inventory, so if an item is not immediately available when a unit orders it, and the price changes before it is delivered, the unit is charged the new price instead of the amount originally obligated when the order was placed (DoD FMR, 2022b, Chapter 15).

In coordination with the sponsor and using inputs from subject-matter experts and analysis of data from GCSS-A and Federal Logistics Data (FEDLOG),[2] we developed several alternative price and credit policies and a set of evaluation criteria that could be used to compare them. The alternative policies included *born-on pricing* (BOP), which establishes the price at the time that the DLR is ordered, and a *single price* (SP) for DLRs instead of separate prices and credits, as well as other policies to address price changes and reliance on credits. We then used the evaluation criteria to identify the most promising courses of action for the Army.

We analyzed data from GCSS-A and FEDLOG to identify the effects of eliminating credits during FY 2021 on Army customers and to estimate the effects of various policy options on reducing workload associated with price changes and tracking credits. We also examined data from AWCF budget submissions to better understand the causes of the AWCF's cash balance problems and other efforts that were undertaken to restore the cash balance, in addition to the credit shutoff.

Outline of This Report

In Chapter 2, we describe the Army's current system for managing purchases of DLRs and consumables and set it within the context of the other services and similar private-sector systems. We also summarize the regulations and policies that govern WCFs and the perspectives of various stakeholders. In Chapter 3, we review the causes of the AWCF cash difficulties and the impacts of the credit shutoff on Army customers across commands and unit types. We introduce the courses of action (alternative price and credit policies) and the evaluation criteria in Chapter 4. Chapter 5 describes supporting data analysis and the results of our evaluation of the courses of action. Chapter 6 summarizes our findings and conclusions.

---

[2] GCSS-A and FEDLOG datasets were provided to us by Army sponsors for use in this study.

CHAPTER 2

Description of Current System

In this chapter, we provide background information about the Army's current processes for budgeting and funds distribution for DLRs and other resources required for unit training, and purchases and returns of DLRs and other repair parts. We also discuss the role of WCFs and the policies and regulations that govern them, as well as the price and credit policies of the other services. We then discuss the perspectives of Army stakeholders that we interviewed and some of the concerns that they raised.

Current Processes

Under the Army's logistics financial management system, units receive budgets to purchase DLRs and other repair parts from the supply system. Figure 2.1 illustrates the budgeting and funds distribution process for spare parts and other resources that units need for train-

FIGURE 2.1

Army Budgeting and Funds Distribution for Depot-Level Reparable Parts

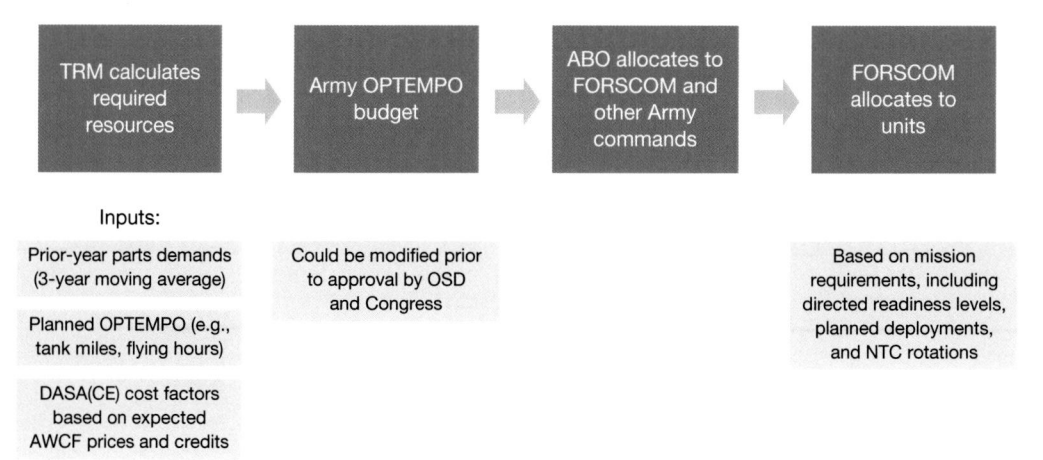

SOURCES: Features information from U.S. Army War College, 2021; U.S. Government Accountability Office, 2007; Army stakeholders, telephone interviews with the authors, 2022.
NOTE: ABO = Army Budget Office; DASA(CE) = Deputy Assistant Secretary of the Army for Cost and Economics; NTC = National Training Center.

ing. The Army uses the Training Resource Model (TRM) to calculate the funds units will need to meet their planned OPTEMPO in terms of vehicle miles or aircraft flying hours. Future demands for spare parts, fuel, and other resources are estimated based on prior-year demands per vehicle mile or flying hour and evaluated based on cost factors developed by the Office of the Deputy Assistant Secretary of the Army for Cost and Economics. These cost factors are based on expected prices and credits for the budget year, which are set as part of the AWCF's budget process.[1]

The operating budgets for Regular Army units are rolled up into the Army OPTEMPO budget as part of the OMA appropriation. The Army's Budget Estimate Submission is reviewed by OSD. When this review process is completed, the Army's OMA budget request becomes part of the President's Budget that is submitted to Congress. Congress may also make changes to the Army budget during the approval process. After the new FY begins and the DoD budget is passed by Congress, the Army Budget Office allocates OPTEMPO funds to FORSCOM and other Army commands. FORSCOM, in turn, allocates funds to units based on mission requirements, including directed readiness levels, planned deployments, and major training events, such as National Training Center rotations.[2]

Figure 2.2 shows the interactions between Army units and the AWCF and other sources of supply. The unit uses its OPTEMPO budget to purchase Army-managed items (AMI, which are primarily DLRs) from the AWCF Supply Management Activity Group and non–Army-managed items (NAMI) from the Defense Logistics Agency (DLA, which manages consumable spare parts) and the General Services Administration (GSA, which manages non-defense-unique supplies and equipment). In some cases, units may also buy DLRs managed by the other services (not shown in the figure). Some commonly used items that have already been purchased from the supply system may be held in the maintenance activity's shop stock.[3]

If the unit is supported by a Supply Support Activity (SSA), the transactions will pass through the SSA. If the item is in the SSA's inventory, the SSA can issue it to the unit immediately.[4] Otherwise, the request will be passed to the source of supply (the AWCF Supply Management Activity Group, DLA, or GSA). These transactions are called 71 purchase orders (POs) in GCSS-A. If the unit is not supported by an SSA, its requests will go directly to the source of supply. These transactions are called 45 POs in GCSS-A.

When a unit orders a DLR, GCSS-A automatically creates a purchase request (PR) for the unit to return the carcass that is removed from the vehicle or other end item when the new DLR is installed. These carcasses are returned to the AWCF Supply Management Activity

---

[1] Based on U.S. Government Accountability Office, 2007, and Army stakeholders, telephone interviews with the authors, 2022. Interviewees are not named to preserve confidentiality.

[2] Based on U.S. Army War College, 2021, Chapter 8, and Army stakeholders, telephone interviews with the authors, 2022.

[3] The description of Figure 2.2 is based on Department of the Army, 2022, and Army stakeholders, telephone interviews with the authors, 2022.

[4] Items in the SSA's inventory are owned by the AWCF.

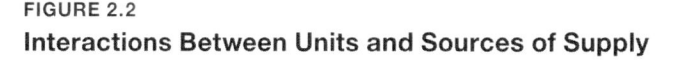

FIGURE 2.2

Interactions Between Units and Sources of Supply

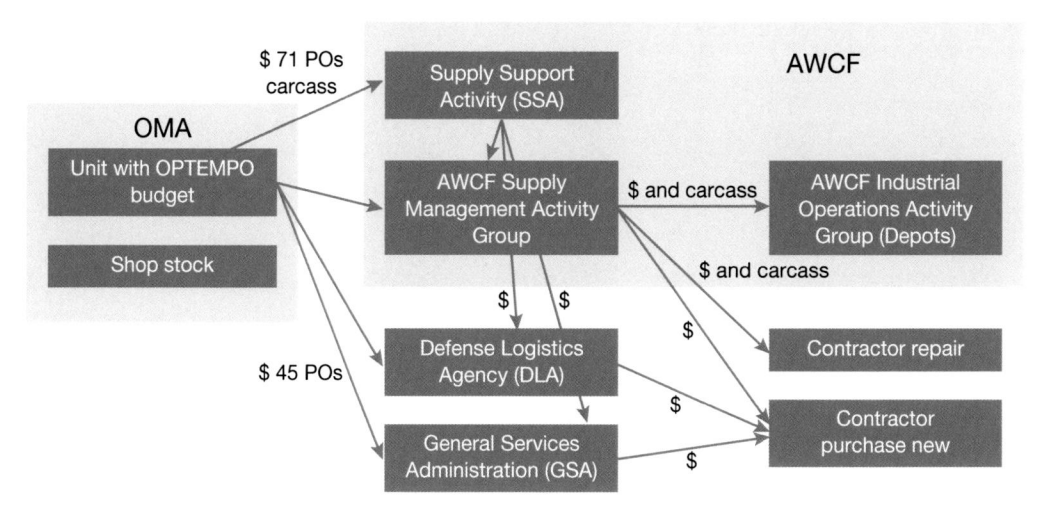

SOURCES: Features information from Department of the Army, 2022; Army stakeholders, telephone interviews with the authors, 2022.

Group, either directly or through the SSA. Under the Army's current price and credit policies, the unit pays the full price for a DLR (latest acquisition cost plus a supply management surcharge to cover operating costs) and receives a credit when the carcass is returned. The AWCF Supply Management Activity Group uses the revenues from sales of DLRs to purchase repairs from organic maintenance depots in the AWCF Industrial Operations Activity Group or from contractors. It can also purchase new DLRs from contractors. Similarly, DLA and GSA use their sales revenues to replenish their inventories with purchases from contractors.

In an unpublished 2021 report on improving retrograde processes for DLRs, RAND researchers James R. Broyles, Kenneth Girardini, Candice Miller, Josh Kerrigan, and Erin Leidy explain how the maintenance and supply systems process DLR orders, issues, and returns. In Figure 2.3, the dark blue boxes at the top of the figure show the high-level steps in the process, and the light blue boxes at the bottom provide additional details. Starting on the left side of the figure, a unit performing field maintenance opens a job order in GCSS-A to initiate a repair. If the mechanic finds that a DLR needs to be replaced, he or she removes the unserviceable DLR and adds a serviceable part reservation to the job order for the replacement DLR. GCSS-A automatically creates an unserviceable part reservation if there is a requirement to return the item. After removing the unserviceable DLR, the unit brings it into their shop stock (Plant 2000) in GCSS-A. The serviceable replacement DLR may be issued quickly if it is available in shop stock or at the supporting SSA, or the unit may have to wait for it to be shipped from the wholesale supply system. GCSS-A automatically creates a return PR for the unserviceable DLR when the serviceable DLR is issued. The return of the unserviceable DLR must be matched with an issue for the unit to receive credit. The unit then issues the unserviceable DLR to their supporting SSA (Plant 2001 in GCSS-A), which may be at the brigade

FIGURE 2.3

Depot-Level Reparable Part Orders, Issues, and Returns

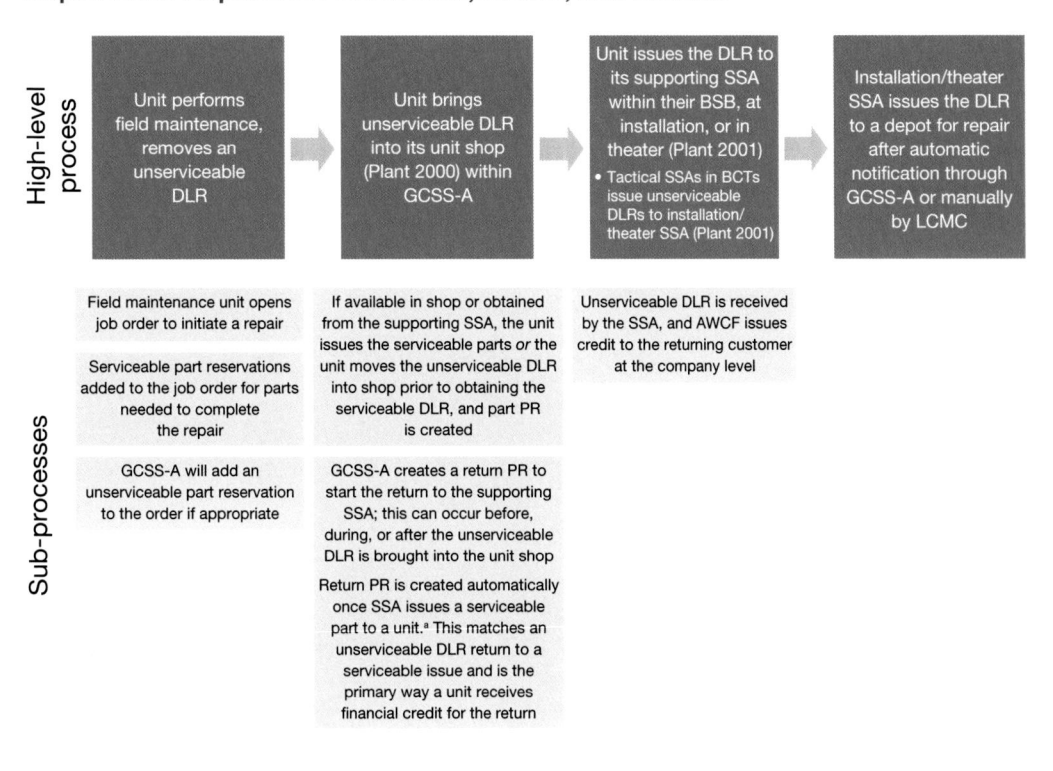

SOURCE: Adapted from unpublished RAND research by James R. Broyles, Kenneth Girardini, Candice Miller, Josh Kerrigan, and Erin Leidy.

NOTE: BCT = brigade combat team; BSB = brigade support battalion; LCMC = Life Cycle Management Command.

[a]The return process is different for excess returns, which are not matched to the issue of a serviceable DLR.

support battalion, the installation, or in theater for deployed units. The financial system then issues a credit from the AWCF to the unit. A tactical SSA in a brigade combat team would return the unserviceable DLR to an installation or theater SSA, and then it is returned to the wholesale supply system for depot-level repair.

Defense Working Capital Funds

The role of defense WCFs, including the AWCF, is to create a buyer-seller relationship between support organizations and their customers, with a goal of promoting more cost-conscious use of resources. For example, units might be tempted to order more DLRs than they immediately need and hold them in shop stock if they do not have to pay for them. Similarly, credits provide a financial incentive to return unserviceable DLRs to the supply system in a timely manner.[5]

WCF-like mechanisms (known as *transfer pricing* in the economics literature) are also used to optimize the allocation of resources within private-sector organizations that have decentralized decisionmaking. Customers within the organization can choose an external provider if it is less expensive than an internal provider, such as a print shop or a business unit making components for another unit's end product. Under these circumstances, prices should be set to provide incentives for internal customers to make decisions that are cost-effective for the organization as a whole. Incentives for cost-effective decisions can be created by using a two-part cost recovery system, where prices reflect the costs that customers impose on the system (marginal or variable costs) and overhead costs are recovered separately from prices through budgets or fixed fees collected from customers. In addition, managers who supervise the internal providers must periodically conduct cost-benefit analyses to determine whether the internal providers have the appropriate capacity and capabilities.[6]

Defense WCFs are intended to perform a similar function as WCF-like mechanisms in private-sector organizations, but they have somewhat different goals and are subject to DoD regulations. In addition to controlling costs, defense WCFs must support operational missions and readiness, so they may need to hold excess capacity or inventories to meet surge requirements. They may also function as a mechanism to allocate scarce resources, such as airlift and sealift capacity provided by U.S. Transportation Command or limited numbers of DLRs in the Army's wholesale supply system.

We reviewed the DoD FMR to identify the sections that currently establish WCF policies, especially those that affect price and credit policies. The DoD FMR indicates that the objectives of WCFs include:

- "Create a cost-conscious environment for both customers and providers."
- "Provide a more effective means for controlling costs . . . and financing, budgeting, and accounting for [those] costs."
- Create contractual relationships between providers and customers.

---

[5] DLRs were brought into the Navy's WCF in the 1980s and the Army and Air Force WCFs in the early 1990s. Bozin (1986) states that one of the reasons the Navy brought aviation DLRs into its WCF was a lack of turn-in accountability for carcasses. There were no procedures to ensure that carcasses were returned when a new DLR was issued and "no real incentive for the squadrons to return carcasses to the system, which caused [unserviceable] parts to remain in squadrons" (p. 41). Byrnes (1993) notes that "DoD budget documents indicate that the Navy customers' annual demand for repairables decreased by 20 percent after conversion to financing repairables through the stock fund" (p. 32). However, WCF pricing policies could also create incentives for inefficient behavior, such as Air Force units purchasing expensive test equipment to determine whether electronic parts removed from aircraft were faulty before returning them to the WCF. (See Jordan, 1995, and Baldwin and Gotz, 1998.)

[6] See, for example, Baldenius and Reichelstein, 2006; Baldwin and Gotz, 1998; Brauner et al., 2000; Byrnes, 1993; and Hirschleifer, 1956.

- Provide WCF managers with "financial authority and flexibility to procure and . . . use manpower, materials, and other resources," including "modern management tools comparable to those used by efficient private enterprises."
- "Facilitate budgeting for and reporting of the costs of end-products."
- "Provide flexibility within budget cycles to changes in supply and demand" (DoD FMR, 2022b, Chapter 1, pp. 1-5–1-6).

The FMR includes several definitions and other provisions related to prices of DLRs and consumable spare parts. The *standard price* is defined as

> The price that customers are charged for DoD Inventory Control Point (ICP)-managed items (excluding subsistence). The standard price remains constant (stabilized price) throughout a fiscal year except for the correction of significant errors. The standard price is computed based on various factors, including the replenishment cost of the item plus surcharges to recover costs for transportation; inventory loss, obsolescence and maintenance; [Capital Investment Recovery]; and supply operations (DoD FMR, 2022a, Chapter 9, p. 9-43).

Elsewhere, the FMR states that "The standard sales price of an item must include recovery of operating costs including payroll, personnel travel, transportation, defense agency billings, other [Defense Working Capital Fund] purchases, operating materials and supplies, rent/communications/utilities, depreciation, transportation and other service contracts" (DoD FMR, 2022b, Chapter 15, p. 15-6). In addition, WCFs must adjust their prices upward or downward to recover prior-year losses or to return prior-year gains to customers (DoD FMR, 2022b, Chapter 11). The *exchange price* is defined as "the price charged to customers exchanging a DLR part that needs repair for a serviceable one (new or repaired). [The exchange price equates] to the latest repair price plus wash out costs (condemnations) per item plus pricing elements necessary to recover other operating costs. (Note: If no carcass . . . is returned, the customer must be charged the full standard price)" (DoD FMR, 2022a, Chapter 9, pp. 9-39–9-40).[7]

In several places, the FMR indicates that prices should be stabilized or fixed during the year of execution. For example, Volume 2B, Chapter 9, of the FMR states, "The [Defense Working Capital Fund] establishes selling prices that are normally stabilized or fixed during execution to mitigate the impact of unforeseen fluctuations that would impact on customers' ability to execute the programs approved by Congress" (DoD FMR, 2022a, Chapter 9, p. 9-6).

---

[7] The Army implemented exchange pricing (EP) in its existing logistics and financial management systems in 2009, but when GCSS-A was fielded beginning in FY 2013, the Army had to make a decision about whether to modify GCSS-A to incorporate EP. At the time, the Army decided to revert to separate prices and credits, but to implement a capability to match carcass returns with issues of DLRs to prevent units from receiving "excess credit" for unmatched returns. See Davis, 2013.

However, there are several exceptions that allow prices to be changed under the following circumstances:

- "To prevent the buildup of excess cash balances or to ensure fund solvency" (DoD FMR 2022a, Chapter 9, p. 9-22)
- To support contingency or emergency operations (DoD FMR 2022a, Chapter 9)
- Correction of significant errors (DoD FMR 2022a, Chapter 9)
- "Unit of issue changes," "first time buys," and "contract modifications" (DoD FMR 2022b, Chapter 15, p. 15-9)
- Discounted prices offered "for items being phased out, . . . with limited shelf life, [in] less than 'fully serviceable' condition, . . . [or] in long supply" (DoD FMR, 2022b, Chapter 15).

WCFs are required to maintain "a positive, daily cash balance at levels exceeding the minimum necessary to meet operating, capital investment, and other justified requirements throughout the year and to support continuing requirements into the subsequent year" (DoD FMR, 2022a, Chapter 9). The AWCF aims to maintain sufficient cash to cover projected disbursements between collection cycles, which occur every seven or eight days, with adjustments for volatility, risk, and cash reserves. For FY 2023, the AWCF had a targeted cash balance between $1.5 billion and $2.8 billion (Department of the Army, 2022). To ensure fund solvency, WCF managers can make out-of-cycle rate adjustments with the approval of the Director, Revolving Funds, within the Office of the OUSD(C) (DoD FMR, 2022a, Chapter 9). Supplemental appropriations can also be used to replenish cash if the balance is negative or approaching negative (DoD FMR, 2022b, Chapter 3), or a supply management WCF may establish a cost recovery element (i.e., increase the surcharge) (DoD FMR, 2022b, Chapter 15).

WCF Supply Management activity groups receive contract authority, which allows them to incur obligations in advance of appropriations or in anticipation of receipts from customers (DoD FMR, 2022a, Chapter 9). However, a sufficient cash balance with the Treasury must be available to support all WCF cash outlays (DoD FMR, 2022b, Chapter 1). As a result, WCFs can have cash flow problems if revenues from customers are lower than expected and there is not enough cash available to cover costs. The DoD FMR also specifies that the "dollar amount of unfilled customer orders accepted at the previous year's standard price must be adjusted (upon notification to and confirmation from the customer) to reflect the latest standard price when notice of the price change is received" (DoD FMR, 2022b, Chapter 15). This statement precludes DoD-wide implementation of BOP, i.e., fixing the price at the time the order is placed.

The DASD(L) convened an integrated product team during FY 2019, including representatives from OSD, the military services, DLA, Special Operations Command, and GSA, to review how the military services manage unmatched transactions (UMTs) and to consider a proposal from the Army to implement BOP.[8] The integrated product team concluded that

---

[8] A UMT occurs when the amount of funds obligated to purchase an item from the supply system is different from the amount of funds disbursed to pay for the part (i.e., when the price changes between the time

the problem of manually adjusting UMTs was unique to the Army, because the other services automatically process these transactions if the cost difference is less than $2,500, whereas the Army's threshold for automatic processing was $250. The other services also set aside prior-year funding to cover these cost increases. Therefore, the team recommended that DoD retain the current policy of revaluing open transactions when prices change at the beginning of each FY, and that the Army increase its threshold for automatic processing to $2,500.[9]

The DoD FMR mentions credits for returning DLR carcasses, but only in the context of depot maintenance. It states that if the depot receives credit from the supply activity for returning a carcass, it must credit the job order for the amount received (DoD FMR, 2022b, Chapter 13). However, we did not find any statement indicating that supply activities are required to give credit for either serviceable or unserviceable DLR returns.

Other Services' Price and Credit Policies

We examined the price and credit policies of the other services to identify options that might be considered by the Army.

Air Force

Currently, the Air Force has different price and credit policies based on the type of item (aviation or nonaviation) and the type of customer (Air Force or other military services). For Air Force customers purchasing nonaviation DLRs, the AFWCF uses EP, which is equivalent to repair costs plus washout costs and a surcharge to cover operating costs (and is also equivalent to the standard price minus the credit). If Air Force customers do not return DLR carcasses within a fixed period of time (such as 60 days), then the AFWCF charges them a penalty equal to the credit for unserviceable items, so that the customer will have paid the exchange price plus the credit, which equals the standard price for the item. When Air Force customers purchase nonaviation consumables (managed by the AFWCF, DLA, or GSA), they pay the effective price at the time the item is issued (i.e., they are not protected from price changes).[10]

For non–Air Force customers purchasing either aviation or nonaviation DLRs, the AFWCF charges the standard price (latest acquisition cost plus a surcharge) and issues credits for unserviceable carcasses that are returned to the AFWCF.

---

the customer ordered the item and the time it was received).

The findings of the integrated product team and its supporting analysis are documented in Lang and Kader, 2020.

[9] The supporting analysis indicated that increasing the threshold to $2,500 would account for 85 percent of the Army's UMTs in FY 2018. The integrated product team also found that implementing DoD-wide BOP would require significant changes to automated logistics and financial systems in the military services and DLA (Lang and Kader, 2020).

[10] This section is based on Light et al., 2018, Appendix B.

For Air Force customers purchasing aviation DLRs and consumables, the AFWCF uses Centralized Asset Management (CAM), which was implemented in FY 2007.[11] The budget for aviation spare parts is managed centrally by the CAM office, which is part of Air Force Materiel Command. The CAM office programs, budgets, and executes Operations and Maintenance (O&M) funds for aviation DLRs and consumables and makes payments to the AFWCF based on the number of flying hours executed by active-duty operating units by aircraft type.[12] As part of that process, the Air Force determines the number of flying hours needed to train aircrews to safely operate aircraft and maintain combat readiness and the projected cost per flying hour (CPFH). The flying hour requirement and CPFH are calculated by aircraft mission design series based on force structure, types and number of aircrew and their training needs by position, and aircraft age. Historical demands for DLRs and consumables and future-year exchange prices (based on approved acquisition and repair costs and surcharges) are inputs to the CPFH calculation. Each month, the CAM office transfers O&M funds to the AFWCF based on the number of flying hours executed by units, multiplied by a CPFH rate.

Unit-level commanders do not receive funds for aviation DLRs or consumables, but they do receive O&M funds for nonaviation DLRs and consumables. Because the costs of aviation DLRs are excluded, unit-level financial management staff manage much smaller O&M budgets than is typical for Army units. In addition, aviation DLRs and consumables are essentially free issue to operating units, and there is no longer a financial incentive to return DLR carcasses. By centralizing financial management at Air Force Materiel Command under the CAM program, the Air Force need for unit-level financial management personnel decreased, resulting in a corresponding manpower decrease. Additionally, price changes have little unit-level impact because any price changes on aviation DLRs and consumables are absorbed in the CPFH rate; nonaviation components are typically less expensive, so price changes have a relatively small effect on unit O&M budgets.

Interviews with Air Force SAF/FMB staff about the financial incentives of the CPFH program indicated that it has not resulted in excessive parts orders or a drop in carcass returns. Under the CAM program, the AFWCF owns all inventory down to the flight line, which may reduce some risk of units trying to hold excess inventory. However, the risk to AFWCF cash balances has increased because of potential mismatches between flying hours and repairs (e.g, if a fleet is grounded, units perform maintenance and order parts from the AFWCF, but the AFWCF receives no revenue as there are no flying hours).

The AFWCF experienced cash flow problems beginning in FY 2019 because the Air Force flying hour program was executed below plan, depot maintenance orders were delayed or deferred, and depot maintenance revenues were affected by production and parts issues. The AFWCF incurred nearly $1.5 billion in net operating losses, and its cash balance dropped by nearly $1 billion, net of a $234 million reprogramming action. Revenues continued to

---

[11] Prior to FY 2007, the AFWCF used EP for Air Force customers purchasing aviation DLRs.

[12] The Air National Guard, Air Force Reserve Command, and Air Force Special Operations Command use CAM processes but manage their own requirements and execution of funds. See Light et al., 2018, p. 85.

be lower than expected in FY 2020 because of the COVID-19 pandemic. To restore its cash balance, the AFWCF slowed spending, increased prices by 8 percent in FY 2020 and nearly 10 percent in FY 2021, and imposed a temporary surcharge of over 5 percent on flying hour customers to stabilize cash. In addition, the AFWCF received cash infusions of $756 million in FY 2020, including appropriated funds and a reprogramming action.[13]

Navy and Marine Corps

The Navy and Marine Corps use EP (which they call a two-price system) for DLRs. The customer pays the exchange price (or net price) when an unserviceable DLR is turned in, or the standard price if a carcass is not returned to the supply system. Retail stock levels (fixed allowances) of DLRs are managed centrally by the Navy. Customers must obtain approval for increases to fixed allowances prior to requisitioning items from the supply system. DLRs are managed under a one-for-one reorder policy; requisitions are limited to a quantity of one each, assuming that an unserviceable carcass will be returned (Department of the Navy, 2017, pp. 2–3).

The Navy funds its DLRs and shipyard depot-level maintenance differently than the other services. Shipyard maintenance was taken out of the Navy Working Capital Fund and switched to mission funding that comes from the numbered fleets. Because most ship-related DLRs are purchased while in port for a maintenance availability, customers on ships do not need large O&M budgets to fund DLRs. In addition, expensive items, such as aircraft engines, are funded with procurement dollars. The Navy also delegates much of its procurement of DLRs and management of consumables to DLA. The Navy determines procurement requirements for DLRs, but DLA purchases them from commercial suppliers. DLA is also responsible for the procurement, management, and storage of consumables based on consumption information the Navy provides (Peltz et al., 2014, Appendix C).

Stakeholder Perspectives

In the remainder of this chapter, we discuss some perspectives on price and credit policies that we heard during our discussions with various stakeholders, including personnel from FORSCOM, III Corps G8, OUSD(C), DASD(L), and the team implementing EBS-C. We conducted these discussions between December 2021 and September 2022; topics included the problems caused by price and credit changes and potential constraints to changing price and credit policies, such as policy constraints and required changes to ERP systems.

---

[13] See U.S. Air Force, 2020; U.S. Air Force, 2021; and U.S. Air Force, 2022.

U.S. Army Forces Command

FORSCOM representatives discussed the effects of changes in prices and reliance on credits on unit budgets and purchasing power. The credit shutoff had a negative effect on the purchasing power of FORSCOM units. FORSCOM managers had to seek additional funds for units to continue their training programs, or units had to reduce training to avoid the cost of purchasing parts for maintenance. Because unit leaders could not be certain whether additional funds would be available, it became more difficult to plan for training. Even under normal circumstances, when credits are issued for carcass returns, receipt of credits can cause units to execute their budgets unevenly, because funds are credited back into their operating budgets.

Personnel from III Corps G8 indicated that if a carcass return is matched with an issue, the unit will usually get credit within a few days. However, there can be delays during end-of-month or end-of-FY closeout, and the transaction may have to be processed manually if the unit's line of accounting changes at the beginning of the FY. The credit automatically goes back to the company level based on the DoD Activity Address Code associated with the return. The credits are issued in current-year dollars, even if the purchase occurred in the previous FY. For this reason, units may delay returning carcasses near the end of the FY so that they will receive funds that can be spent during the new FY.[14] More generally, units may hold on to carcasses if they do not have a matching issue, because unmatched returns do not receive credit.[15] However, this behavior is difficult to track because the carcasses are not typically brought into inventory records until the unit is ready to return them.

When nearly all prices change at the end of a FY, there can be many outstanding transactions that have not yet been filled that are revalued at the new prices. Personnel in units or FORSCOM G8 must either approve the orders at the new prices or deobligate the funding. In addition, prior-year funding must be available to pay for any price increases. FORSCOM financial managers have been reluctant to hold back enough prior-year funding to cover anticipated price increases because of the risk that some orders may be cancelled or backorders may not be filled for more than a year, so the prior-year funding cannot be spent. As a result, manual processes are required to reconcile UMTs and to make sure that sufficient prior-year funding is available in the correct line of accounting. FORSCOM personnel indicated that each UMT takes about half an hour to process and that 4,000 UMTs are equivalent to an entire person-year of effort.[16]

However, according to III Corps G8 personnel, price changes do not have much direct effect on operating units. The parts will be delivered even if the funding is not yet avail-

---

[14] We are able to observe reductions in the value of DLR carcasses returned near the end of each FY in the GCSS-A data we analyzed to create Figures 5.4 through 5.6.

[15] This may occur, for example, if an expensive DLR is issued from shop stock to complete a repair but the unit decides not to replenish the shop stock immediately.

[16] GCSS-A data indicate that there were about 200,000 UMTs with price increases over three FY boundaries from FY 2019 to FY 2022, or about 67,000 each year.

able, and the UMTs are handled at the division, corps, or FORSCOM level. The unreconciled transactions will remain open until sufficient prior-year funds are available and the UMT can be resolved.

Office of the Under Secretary of Defense, Comptroller

Representatives of OUSD(C) indicated that any changes to Defense Working Capital Fund policies in the DoD FMR would require approval by the other military services and the Office of the Under Secretary of Defense for Acquisition and Sustainment. This requirement affects BOP because the FMR specifies that the supplier can charge the price that is in effect when the item drops from inventory. The DoD FMR would not prevent the Army from absorbing price changes in the AWCF, but the Army would need to forecast the amount of additional cash that would be needed (net of any price decreases) to implement the policy change and raise the supply management surcharge to account for it. This type of policy change might be riskier in a period of high inflation.

In addition, OUSD(C) would have concerns about a policy change, such as SP, which would remove financial incentives to return DLR carcasses, because the carcasses are needed to feed into depot repair programs. These repair programs reduce overall support costs because they are less expensive than buying additional new DLRs. The Army would still need to be able to monitor carcass returns to ensure that they were comparable with the number of DLRs issued, including DLR returns by non-Army customers. Furthermore, customers in the other services who use Army-managed DLRs would expect to receive credit for carcass returns if they are required to pay the standard price for DLR issues.

Enterprise Business System Convergence

Members of the Army team that is planning for EBS-C provided information about the feasibility of price and credit policy changes within legacy logistics and financial management systems and the implementation timeline for EBS-C. Under EBS-C, the Army will merge GCSS-A, AMC's wholesale-level Logistics Modernization Program (LMP), and the General Fund Enterprise Business System (GFEBS) into a single enterprise resource planning (ERP) system. EBS-C will also take over the role of the Army Enterprise Systems Integration Program and other business systems. Software company SAP SE decided to phase out prior versions of its software that were used to implement GCSS-A, LMP, and GFEBS, and the Army determined that it would be better to combine the three systems than to upgrade each system individually. The EBS-C team is currently working with Army stakeholders to decide what business processes it will implement in the new system, including price and credit policies. Initial partial deployment is planned for FY 2025, and full deployment is planned for no later than FY 2032.

Some price and credit policy changes that would be difficult to implement in three legacy systems would be easier in a single ERP. EBS-C representatives indicated that implementing BOP in the new system would be relatively easy. However, it would be difficult to absorb all price changes in the AWCF, because some transactions go directly to DLA and GSA and do not pass through the AWCF. Moving the point of sale to the job order is considered an industry best practice, because it gives the supply system visibility when parts are used for repair and when an unserviceable DLR should be returned.[17] Matching purchases with turn-ins, which would be required under the Army's current price and credit policies and for EP, would be difficult. However, removing the current system interfaces and giving real-time visibility of parts consumption could make matching returns to issues less important.

---

[17] In the Army context, this would require moving OMA-funded inventories currently held in GCSS-A Plant 2000 (including shop stock) into the AWCF.

CHAPTER 3

Historical Impact of Price Changes on U.S. Army Forces Command

In this chapter, we describe the causes of AWCF cash flow problems that preceded the credit shutoff in FY 2021 and the financial effects of the credit shutoff on FORSCOM, other major Army commands, and different types of units.

Causes of Army Working Capital Fund Cash Flow Problems

Figure 3.1 shows AWCF sales, credits, cash balance, and contract authority from FY 2011 through the FY 2023 President's Budget. At the beginning of the period, sales were falling as operations in Iraq and Afghanistan were drawing down. Credits were low because the AWCF was not issuing credits to units in the U.S. Central Command area of operations. Contract authority, which allows the AWCF to purchase inventory ahead of sales, was much lower than sales, which indicates that the AWCF was drawing down inventory and accumulating cash.[1] Contract authority remained substantially below sales through FY 2015, and eventually, the inventory drawdown resulted in supply availability problems. Beginning in FY 2016, contract authority was increased to match net sales and then rose above net sales in FY 2018 to FY 2020, which increased inventory and reduced available cash.

When COVID-19 restrictions began in March 2020, they had a significant impact on the AWCF cash balance because of reduced training and purchases of DLRs. The Army estimated that total operational cash losses associated with COVID-19 exceeded $1.6 billion (Department of the Army, 2021). Figure 3.2 shows the various actions that the Army took to restore the cash balance, in addition to the credit shutoff. The AWCF received cash infusions of $492 million in September 2020 and $920 million in FY 2021 and requested a direct appropriation of $323 million in FY 2022 to offset losses.[2] Contract authority was reduced below sales in FY 2021 and FY 2022 (shown in Figure 3.1). The supply management surcharge increased from

---

[1]   Much of this cash was transferred out of the AWCF in FY 2011 and FY 2012, as shown in Figure 3.2.

[2]   Because of the timeline of the budget process, by the time the AWCF received this appropriated funding in FY 2022, its cash balance was already recovering. However, the cash balance in FY 2022 would have been about $1.91 billion without this additional funding.

FIGURE 3.1

Army Working Capital Fund Sales, Credits, Cash Balance, and Contract Authority

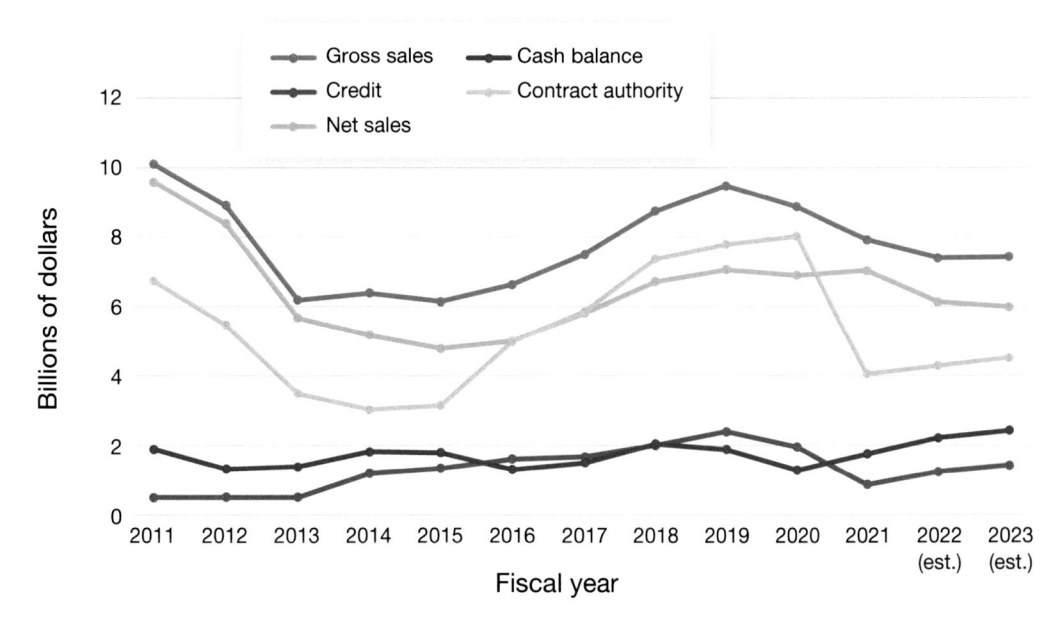

SOURCE: Features information from Department of the Army, 2012–2022.

11 percent in FY 2020 to 15.6 percent in FY 2021 and 23.4 percent in FY 2022, leveling off at 22.9 percent in FY 2023. In addition, the unit cost (a measure of total costs divided by sales) was reduced from 1.23 in FY 2020 to 0.76 in FY 2021, but then increased to 0.9 in FY 2022 and 0.95 in FY 2023.[3] These actions restored the cash balance to its normal range in FY 2022.

Effects of Credit Shutoff on Army Commands and Units

The credit shutoff that occurred during FY 2021 was much more dramatic than previous mid-year price and credit changes experienced by Army customers. Figure 3.3 shows the magnitude of mid-year price changes.[4] In addition, we show the number of serviceable and

---

[3] Unit cost is defined as (obligations + credit + depreciation expense) / (gross sales). A unit cost above 1.0 indicates that the Army is purchasing inventory in anticipation of future needs. A unit cost below 1.0 indicates that the AWCF is reducing inventory by selling and not replenishing spare parts. See Department of the Army, 2021, pp. 27–28.

[4] Price changes occurring in October are omitted from the chart, because almost all prices change at the beginning of each FY, when surcharges are adjusted. The magnitude of these changes would dominate mid-year price changes.

FIGURE 3.2

Actions Affecting Army Working Capital Fund Cash Balances

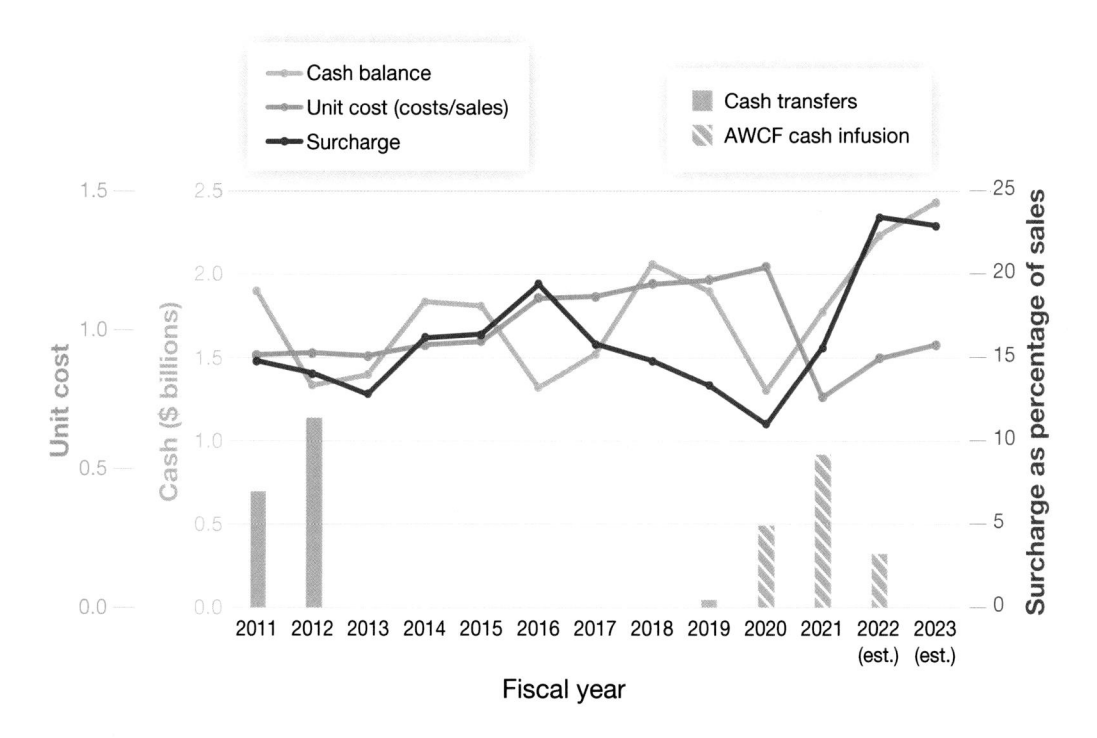

SOURCE: Features information from Department of the Army, 2012–2022.

unserviceable credits turned on or off in each FY. In FY 2019 and FY 2020, the AWCF turned off unserviceable credits for a larger number of DLRs than in prior years, but these changes were dwarfed by the shutoffs of both serviceable and unserviceable credits in FY 2021. In the remainder of this section, we will discuss the financial impacts of these credit shutoffs on Army commands and units.

To estimate the impacts of the FY 2021 credit changes on Army commands and units, we created monthly snapshots of prices and credits from the FEDLOG catalog on the first of each month.[5] From GCSS-A, we pulled all customer turn-ins (movement types 901 and 907 in the MSEG table). We then estimated actual credits received by units based on the credits in effect on the first of each month. So, for example, all customer turn-ins that occurred in May 2021 were evaluated at the credits in effect on May 1, 2021. To estimate how much credit was reduced, we evaluated the same customer turn-in transactions using the credits in effect on

---

[5] In practice, price and credit changes can be posted at any time of the month, so using monthly snapshots is an approximation of the actual prices paid and credits received by customers. However, most price and credit changes are entered at the beginning of each month.

FIGURE 3.3

Changes in Prices and Credits for Depot-Level Reparable Parts

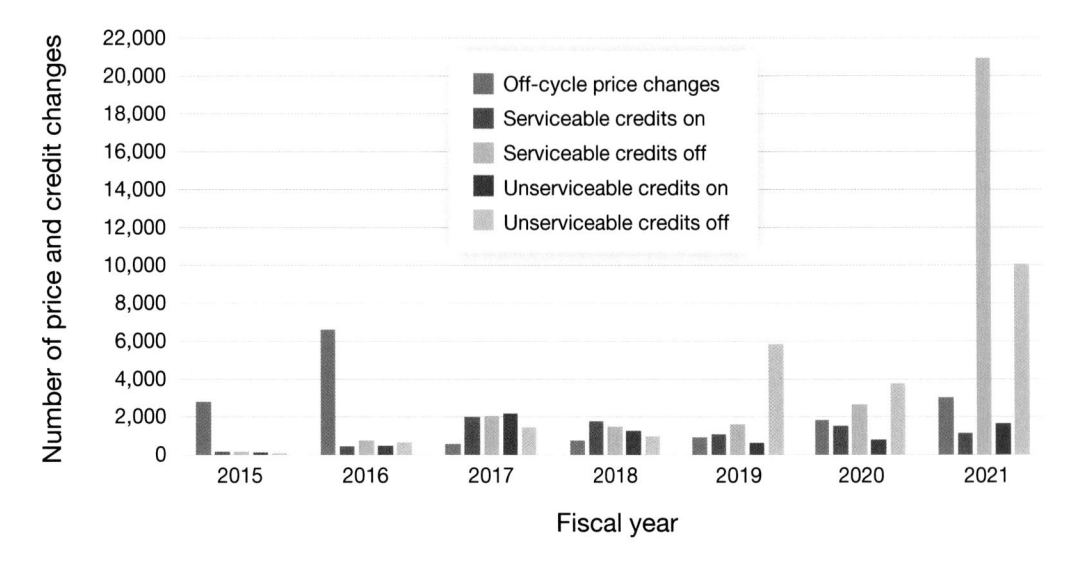

SOURCE: Features monthly FEDLOG data.
NOTE: This chart omits 49,000 price changes in February 2016 and credits that were turned on in July 2021 and off in August 2021, which may have been catalog errors.

January 1, 2021, before the AWCF shut off credits. The credit reductions first appeared in the monthly catalog snapshots on March 1, 2021.

We show estimated Army-wide reductions in the credits received by units during calendar year 2021 in Figure 3.4. The green portions of each column show the estimated value of serviceable and unserviceable credits that units continued to receive, and the red portions of each column show the value of the credits that units would have received if credits had remained the same as in January 2021. At first, all credits were shut off in March 2021, but credits for some high-demand items were restored in April 2021. In July 2021, most credits were restored, but then shut off again in August 2021. Most credits were reinstated at the beginning of FY 2022 (October 2021). Estimated Army-wide reductions in credit received by units for calendar year 2021 are $700 million in unserviceable credit and $157 million in serviceable credit. The reduction in credit affected units' buying power, because this funding (which was anticipated when OMA funding was budgeted) was not available to purchase DLRs and other resources needed for training.

The distribution of credit reductions across major Army commands is shown in Figure 3.5. FORSCOM units had the largest estimated reductions, amounting to $307 million, followed by AMC, the Army National Guard (ARNG), and U.S. Army Pacific. Monthly estimated credit reductions in calendar year 2021 for FORSCOM and Army-wide are shown in Table 3.1. Most of the estimated credit reductions occurred between March 2021 and September 2021

FIGURE 3.4

Army-Wide Financial Impacts of 2021 Credit Shutoff

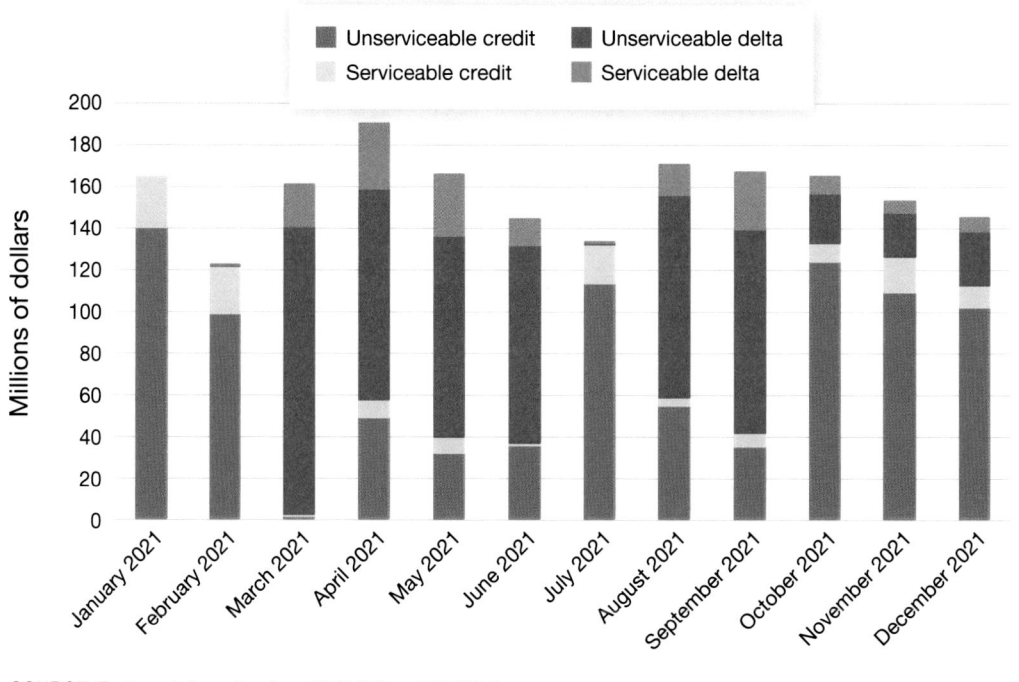

SOURCE: Features information from FEDLOG and GCSS-A.

($262.3 million for FORSCOM), although relative to January 2021, the AWCF was still not issuing credits for some DLRs in the first three months of FY 2022.

The credit shutoff also had varying impacts on different unit types, as shown in Figure 3.6. Units that typically purchase larger numbers of expensive DLRs, such as armored brigade combat teams (ABCTs) and combat aviation brigades (CABs), incurred much larger credit reductions during calendar year 2021 than infantry brigade combat teams (IBCTs), Stryker brigade combat teams (SBCTs), and sustainment brigades. The "other" category also includes some types of units with expensive equipment, such as combat engineer, field artillery, and aviation units.

We also observed some variation in credit reductions within unit types, depending on the unit's directed readiness level, planned deployments, and National Training Center rotations. Figure 3.7 shows these variations for selected ABCTs, CABs, IBCTs, and SBCTs. Note that the y-axis scales vary across some of the panels of the figure.

The unplanned reduction in credits can affect Army units' ability to train, because the credits flow back into their OMA budgets and can be used to purchase additional spare parts or other resources required for training. Changes to price and credit policies could potentially reduce Army units' reliance on credits, but they would not prevent the AWCF from getting into cash difficulties. If shutting off credits were no longer an option, the Army would need

FIGURE 3.5

Financial Impacts of 2021 Credit Shutoff by Major Command

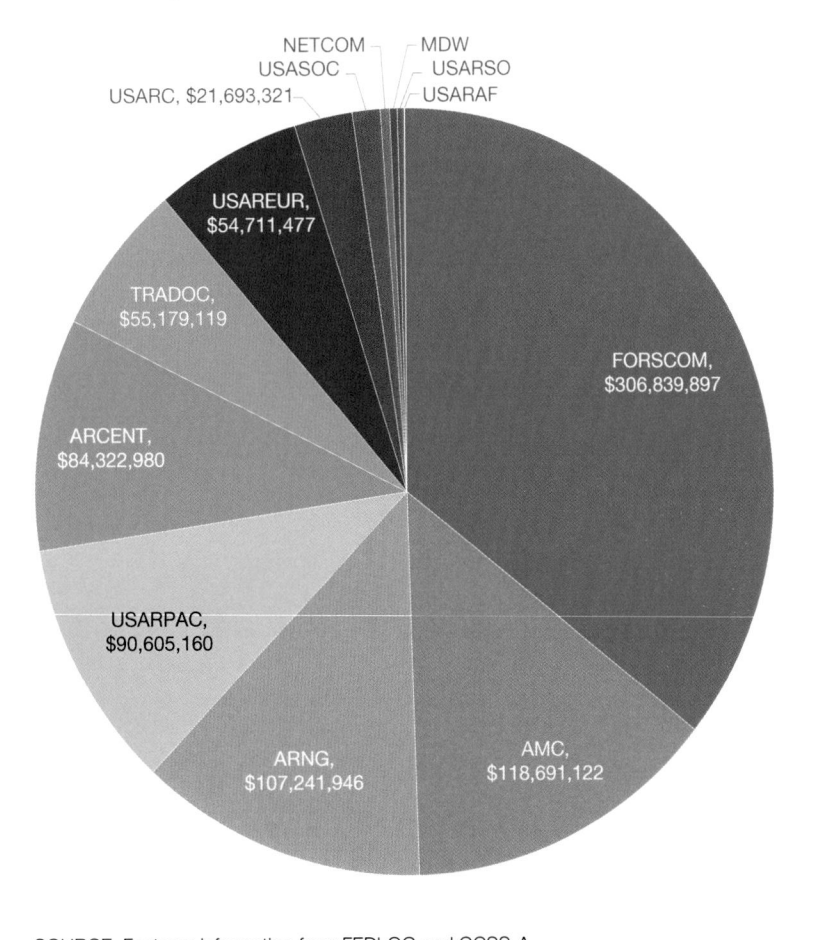

SOURCE: Features information from FEDLOG and GCSS-A.

NOTE: ARCENT = U.S. Army Central, TRADOC = Training and Doctrine Command, USAREUR = U.S. Army Europe, USARC = U.S. Army Reserve Command, USASOC = U.S. Army Special Operations Command, NETCOM = U.S. Army Network Enterprise Technology Command, MDW = Military District of Washington, USARSO = U.S. Army South, USARAF = U.S. Army Africa.

to use different policy levers, such as cash infusions, reductions in purchases, and increases in surcharges to restore the AWCF cash balance. These actions could also affect supply availability and reduce the buying power or the amount of OMA funding available to Army units.

TABLE 3.1

Monthly 2021 Credit Reductions for U.S. Army Forces Command and Army-Wide

Month	FORSCOM		Army-Wide	
	Unserviceable Credit Reductions	Serviceable Credit Reductions	Unserviceable Credit Reductions	Serviceable Credit Reductions
January	$0	$0	$0	$0
February	$105,495	– $14,413	$447,684	$56,752
March	$55,127,660	$5,090,404	$139,027,996	$20,432,714
April	$35,228,510	$6,075,043	$101,640,984	$31,482,442
May	$33,130,563	$9,564,210	$97,214,535	$29,702,457
June	$32,879,217	$3,956,678	$95,434,702	$12,834,614
July	$154,181	$59,087	$818,833	$59,425
August	$33,463,250	$3,774,559	$98,016,739	$14,521,246
September	$38,310,535	$5,500,418	$98,261,409	$27,779,367
October	$12,330,799	$2,727,785	$23,977,943	$7,987,641
November	$12,487,082	$2,092,949	$21,336,215	$5,448,363
December	$12,474,538	$2,271,004	$25,857,393	$6,547,733
Total	$265,691,830	$41,097,724	$702,034,432	$156,852,752

SOURCE: Features information from FEDLOG and GCSS-A.

FIGURE 3.6

Impacts of Credit Reductions by Unit Type, Calendar Year 2021

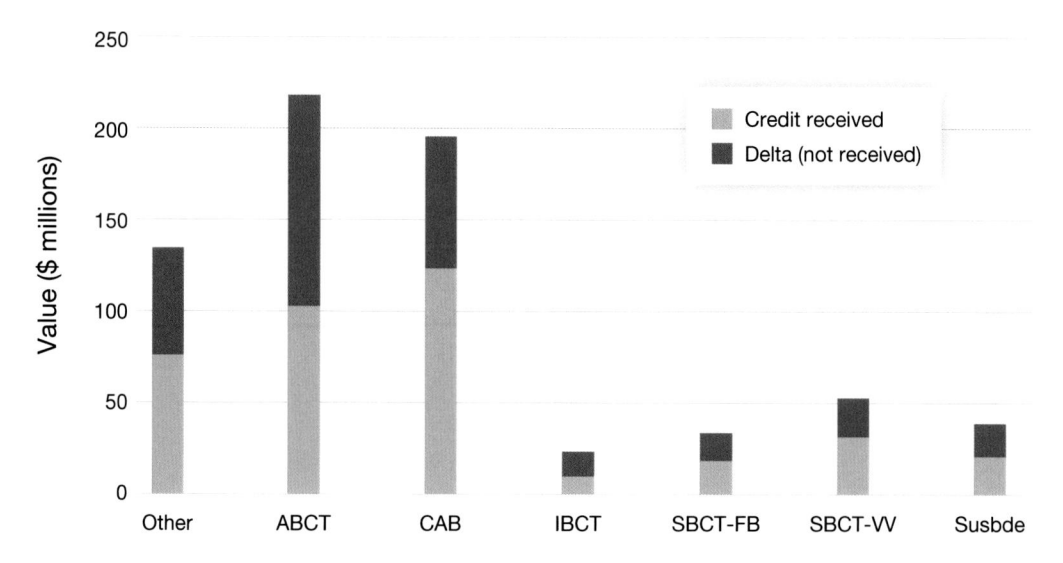

SOURCE: Features infromation from FEDLOG and GCSS-A.
NOTE: FB = flat bottom; susbde = sustainment brigade; VV = double-V hull.

FIGURE 3.7
Impacts of Credit Reductions Within Unit Types, Calendar Year 2021

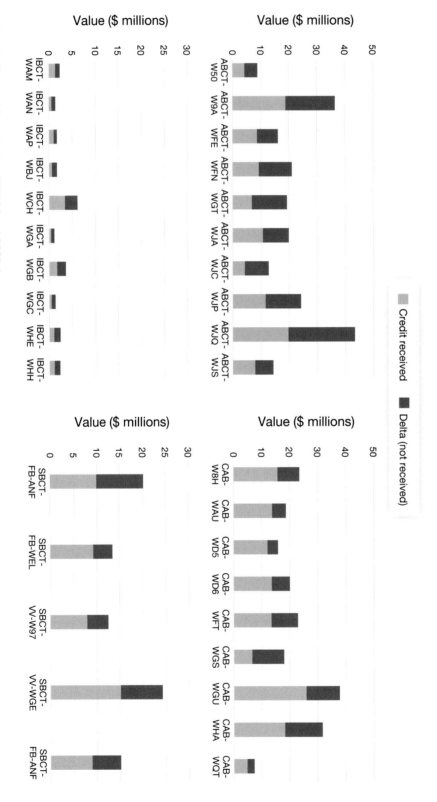

SOURCE: Features information from FEDLOG and GCSS-A.

CHAPTER 4

Courses of Action and Evaluation Criteria

In this chapter, we discuss the courses of action that the Army could use to reduce the effects of price changes and reliance on credits. We also introduce the evaluation criteria that we used to evaluate the alternative courses of action.

Courses of Action

We developed, in consultation with the sponsor, eight courses of action to evaluate. These courses of action are intended to address two distinct sets of problems that were raised during our discussions with Army stakeholders. The first set of problems is related to price changes between the time a requisition is placed and the time the order is received. These price changes generate financial workload, particularly if there are not sufficient funds in the customer's OMA account to pay for a price increase. Although there are some price changes that occur within an FY, they are most significant at the end of each FY, when virtually every outstanding transaction is revalued at the prices for the new FY. To resolve the UMTs that are created by these FY price changes, the customer must have sufficient prior-year OMA funds to pay the difference. However, units and commands often try to fully execute OMA funds by the end of the FY on September 30 to avoid losing access to these funds. Therefore, UMTs take time to process because financial managers at the division, corps, and FORSCOM levels must find the prior-year funds and ensure that they are available in the correct line of accounting.

The first four courses of action listed in Table 4.1 are intended to address problems associated with price changes. The first three are variations on the idea of BOP, under which the price is set at the time the customer places an order, not at the time it is delivered. Under BOP A, BOP would be established DoD-wide, for all customers. However, this course of action would require agreement from the other military services and DLA, and it was recently rejected by a DoD-wide integrated product team. Therefore, we also considered variations of BOP that could be implemented by the Army without agreement from the other services. BOP B would absorb price changes in the AWCF for all purchases that pass through the AWCF and would apply to Army customers only.[1] BOP C would apply to AMI only and

---

[1] This approach excludes some types of transactions that go directly to the source of supply and do not pass through the AWCF, as we will discuss in more detail in Chapter 5.

TABLE 4.1

Price and Credit Policy Courses of Action and Intended Effects

Course of Action	Price Changes	Reliance on Credits	Unmatched Returns
BOP A (DoD-wide)	✓		
BOP B (in AWCF)	✓		
BOP C (AMI only)	✓		
RT	✓		
SP		✓	✓
EP		✓	
Moving point of sale to job order (MPS)			✓
Free issue (FI)	✓–	✓	✓

NOTE: A check mark indicates that the course of action addresses the problem. A check minus indicates that it partially addresses the problem.

would include all Army customers, but would exclude purchases from DLA, GSA, and other sources of supply. Finally, the Army could raise the threshold for automatic processing of UMTs (RT) from $250 to $750, which would reduce manual financial management workload, provided that funding is available in the correct line of accounting.

The second set of problems involves financial management workload associated with credits. Our interviewees indicated that units typically receive credits within a few days of returning a carcass, but receipt of credits may cause uneven execution of OMA budgets during the FY, as a unit's spending may slow or even reverse when credits are received, making it more difficult to monitor budget execution and reallocate funding. In addition, the requirement for carcass returns to be matched with a purchase of the same item may cause units to hold onto carcasses until they have a matching issue. As a result, DLR carcasses might not be returned in a timely manner or be available for induction into depot repair programs.

There are four additional courses of action intended to address one or more credit-related problems. SP would set the price of a DLR net of the repair cost (equal to the standard price minus the credit) and eliminate credits. It would rely on supply system discipline to ensure that DLR carcasses are returned for depot repair. It would also eliminate reliance on credits, and there would be no financial incentive for units to keep unmatched returns. An EP system would eliminate unserviceable credits but would not provide credit for unmatched returns. However, it would also maintain a financial incentive to return matched carcasses, because the unit would receive a "delta bill" equal to the unserviceable credit if soldiers failed to return a carcass during a fixed period of time, such as 60 days, after the new DLR was issued.

MPS would essentially involve moving some OMA-funded inventories currently held in Plant 2000 (including shop stock) into the AWCF and making these parts visible at the wholesale level. A DLR taken out of these inventories would then count as an issue, so there would be a one-to-one relationship between issues and returns of unserviceable DLRs taken off end items during the repair process. Thus, it would eliminate unmatched returns, but it would not eliminate reliance on credits unless other changes were made to price and credit policies. We also evaluated an FI policy, under which units would not pay for DLRs and funding would go directly to AMC to purchase and repair DLRs.[2] Prices and credits for DLRs would be eliminated, but units would still be affected by price changes on NAMI.

Evaluation Criteria

In this section, we describe the evaluation criteria that we used to compare the courses of action with the status quo, which we define as separate price and credit transactions for issues and returns of DLRs.[3] We developed these criteria in consultation with the sponsor. We evaluated each course of action based on a combination of data analysis and subject-matter–expert opinion. These criteria could be weighted to emphasize specific factors, such as potential costs to the Army or effects on readiness. However, in practice, we found that the ranking of the courses of action seemed clear without any weighting of the criteria.

- **Effects on unit buying power:** To what extent does the course of action reduce financial uncertainty associated with price changes or improve the steady execution of unit budgets by reducing reliance on credits?
- **Effects on financial management workload:** Does the course of action reduce workload associated with reconciling UMTs or monitoring credits and unmatched returns?
- **Effects on DLR demands and carcass returns:** Does the course of action provide financial incentives or other mechanisms to limit customer demands for DLRs and encourage returns of unserviceable DLR carcasses?
- **Effects on financial stability of the AWCF:** Does the course of action create any additional risk to AWCF solvency or the functioning of the Army's wholesale supply system relative to the status quo of separate prices and credits?

---

[2] This policy was in effect prior to stock funding of DLRs, which was implemented by the Army in 1992. See Pint et al., 2002.

[3] This policy was implemented in FY 2013 when GCSS-Army was fielded. Note that under this policy, each DLR return must be matched with an issue to receive credit, and prices can change if they are updated in the catalog in the time between the order and the receipt of the item.

- **Required approvals by other Army commands, other military services, DoD, or Congress:** Are other Army commands likely to object to the course of action, or would it require approval by the other military services and DoD, or new legislative authority?
- **Required changes to legacy or future ERPs:** How difficult or costly would it be to implement required changes in either existing ERPs or EBS-C? Could the change be implemented in existing systems or would it have to wait until EBS-C is fielded?
- **Required changes to budgeting or funds distribution:** Would the Army need to make changes to the way it currently budgets for DLRs and distributes funding to implement the course of action? Would it require changes to the TRM?
- **Effects on other services and Army reserve components:** How would the course of action affect the other services, the Army Reserve, and the ARNG?

To provide some perspective on potential effects of price and credit policy changes on the Army reserve components and the other services, Table 4.2 provides some recent data on the share of new orders and revenue received by the AWCF supply management activity group from these sources. Although the Regular Army is by far the largest customer of the AWCF supply management activity group, the Army reserve components and other services account for about $1 billion in annual sales.

In the next chapter, we will discuss how we applied these criteria to the courses of action.

TABLE 4.2

Army Working Capital Fund Supply Management Sources of New Orders and Revenue (Millions of Dollars)

Source	FY 2021	FY 2022 (est)	FY 2023 (est)
OMA	$5,004.9	$5,282.4	$5,513.3
OMARNG	$590.2	$667.1	$653.5
OMAR	$99.7	$67.2	$65.7
Navy	$93.7	$104.9	$73.6
Air Force	$100.1	$89.1	$77.3
Marine Corps	$54.4	$61.2	$59.5
Total (including other sources)	$7,469.2	$7,117.9	$7,224.2

SOURCE: Features information from Department of the Army, 2022.

NOTE: OMAR = Operations and Maintenance, Army Reserve; OMARNG = Operations and Maintenance, Army National Guard.

CHAPTER 5

Course of Action Analysis

In this chapter, we first review data analysis we conducted to estimate the potential benefits of policies intended to reduce financial management workload associated with price changes and to assess the effects of removing credits as an incentive to return unserviceable DLR carcasses. Next, we discuss two examples of the approach we used to evaluate each course of action and summarize the results for each course of action.

Data Analysis

Policies Addressing Price Changes

There are two types of policies intended to address workload associated with price changes. One type consists of variations on BOP (fixing the price at the time the part is ordered), and the other involves raising the threshold for automatic processing of UMTs. To understand the extent to which these courses of action would reduce the financial management workload associated with UMTs, we analyzed data from GCSS-A and FEDLOG over the period from October 2018 to June 2022, which covers three FY transitions. The transaction data from GCSS-A include POs for AMI and NAMI Class IX (spare parts). We used archived monthly catalog snapshots (on the first of each month) from FEDLOG to evaluate whether the price when each item was issued was different than when the item was ordered. For 71 POs (which pass through an SSA), we compared the price on the DoD document date with the SSA issue date,[1] and for 45 POs (which do not pass through an SSA), we compared the price on the DoD document date with the shop receipt date.[2] If the PO and the receipt transaction are in the same month, we assume there was no price change.

Figures 5.1 and 5.2 indicate that the vast majority of price changes occur at the beginning of each FY for both NAMI and AMI. As we discussed in Chapter 2, these price changes create financial management workload at the division, corps, and FORSCOM level, whereby personnel must process the UMTs and make sure there is sufficient prior-year funding available

---

[1] The SSA issue date is defined as a movement type 643 (Plant 2001) in the MSEG table.

[2] The shop receipt date is defined as a movement type 101 (Plant 2000) in the MSEG table.

FIGURE 5.1

Monthly Price Changes for Non-Army-Managed Items

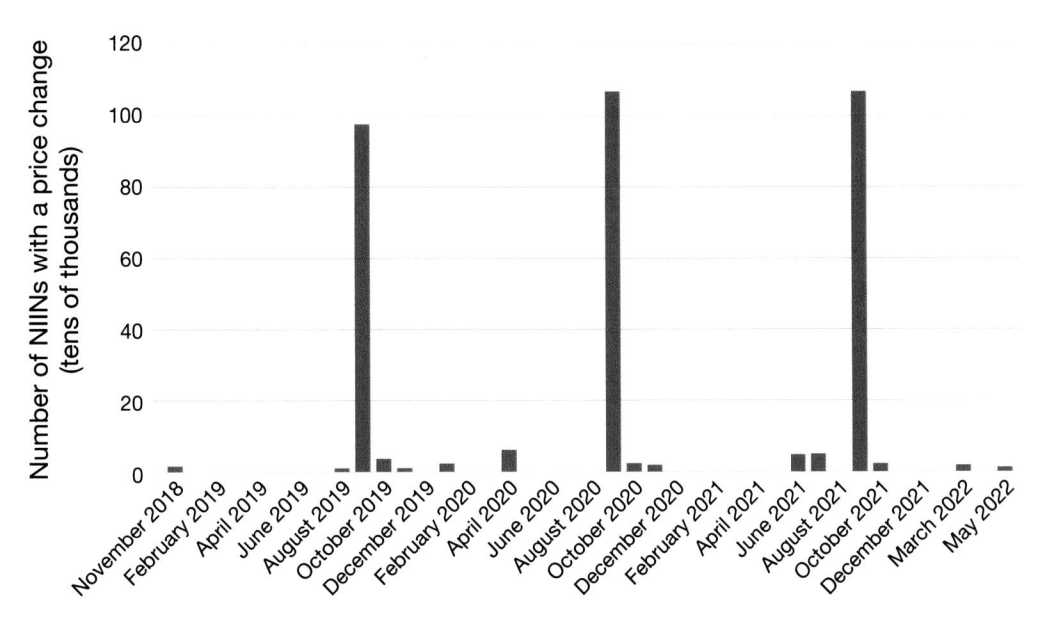

SOURCE: Features information from monthly FEDLOG data extracts.
NOTE: NIIN – National Item Identification Number.

FIGURE 5.2

Monthly Price Changes for Army-Managed Items

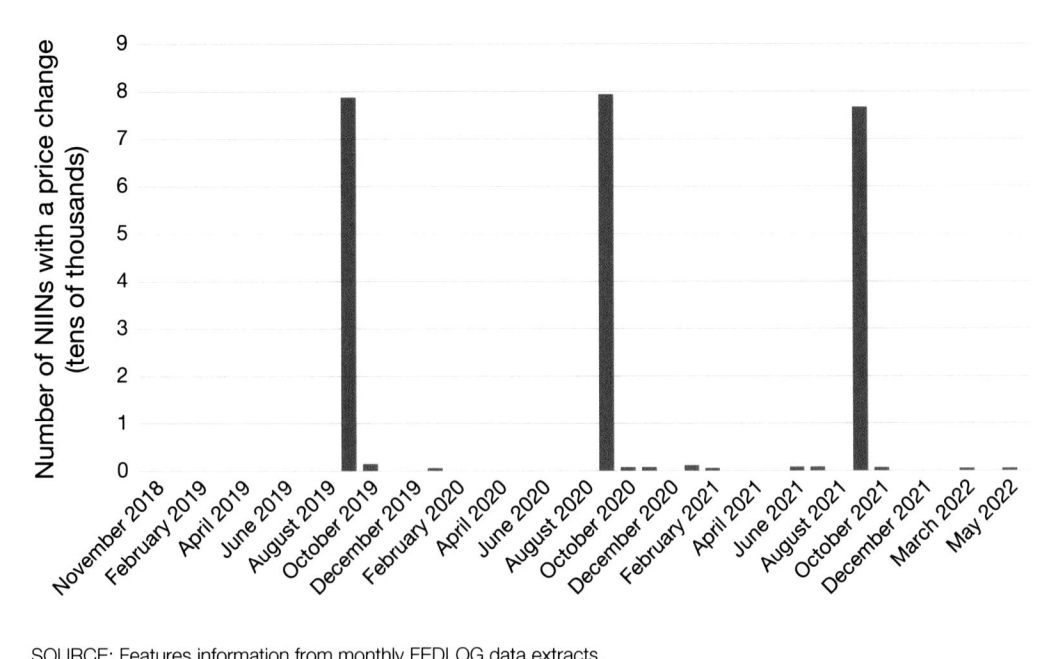

SOURCE: Features information from monthly FEDLOG data extracts.

in the correct line of accounting to pay for increased prices. Although parts are delivered as they become available, the UMTs persist in GFEBS until they are resolved.

Table 5.1 shows the estimated number of price changes that occurred on items ordered and received by FORSCOM units during the period from October 2018 to June 2022, along with the dollar value of those changes. Only about 33,000 of these price changes occurred in the same FY, with a net value of about $3.5 million, because about 20 percent were price decreases. Of the 290,000 transactions that crossed an FY boundary, 92 percent had either a price increase or decrease. Nearly 200,000 had price increases, with a net value of over $100 million over three FY transitions. We assume that only price increases generate financial management workload, because price decreases would result in a credit to the original line of accounting. Price increases accounted for approximately 215,000 total transactions over the period, with a net value of about $110 million.

Next, we estimated how much of the UMT workload each course of action would eliminate. If the Army set the threshold for automatic processing of UMTs at $750 or less, that would account for all but 20,500 transactions, or a 90 percent reduction. The dollar value of these price changes is about $13 million, which is less than the value of the price decreases. Because $750 might be a relatively small change in the price of an expensive DLR, we also examined the effect of adding a requirement that the price change must be greater than 20 percent of the unit price. Combining these two criteria reduces the number of remaining transactions to 13,000, or a 94 percent reduction. The total dollar value of this larger set of price changes is $41 million, because it included higher dollar-value changes on expensive items. However, there might still be some financial management workload involved in making sure that the required additional prior-year funding is available in the correct line of accounting.

We also looked across all Army commands to examine the effects of varying the dollar threshold for automatic processing of UMTs. These results are shown in Table 5.2. Each panel of the table shows the number of remaining transactions and their dollar value as the threshold is raised. A threshold of $250 accounts for 87 percent of transactions and 5.4 percent of the dollar value of transactions. If the threshold is raised to $750, the share of transactions

TABLE 5.1

U.S. Army Forces Command Transactions with Price Changes

Timing of Price Change	Type of Price Change	Number of POs	Total Change in Value
Same FY	Decrease	6,607	−$3,854,757
Same FY	Increase	16,347	$7,296,589
Different FY	Decrease	66,284	−$29,238,529
Different FY	No Change	25,355	$0
Different FY	Increase	199,490	$135,945,779
Net change in value			$110,149,083

SOURCE: Features information from GCSS-A EKPO and MSEG tables and monthly FEDLOG data extracts.

TABLE 5.2

Effects of Raising the Threshold for Automatic Processing of Unmatched Transactions

Time Period	Type of Price Change	All Transactions		Price Changes > $250		Price Changes > $750		Price Changes > $1,000		Price Changes > $2,500	
		Number of POs	Dollar Value	Number of POs	Dollar Value	Number of POs	Dollar Value	Number of POs	Dollar Value	Number of POs	Dollar Value
Same FY	Decrease	18,307	–$7.6m	2,012	–$7.1m	996	–$6.7m	803	–$6.5m	420	–$5.9m
Same FY	Increase	43,639	$17.5m	5,193	$16.5m	2,533	$15.3m	2,124	$14.9m	1,188	$13.4m
Different FY	Decrease	203,772	–$59.3m	27,684	–$54.2m	12,957	–$47.8m	10,151	–$45.3m	4,359	–$36.4m
Different FY	No change	84,425	$0	0	$0	0	$0	0	$0	0	$0
Different FY	Increase	659,104	$303.2m	87,631	$288.6m	46,774	$270.3m	38,589	$263.2m	18,841	$231.6m

SOURCE: Features information from GCSS-A EKPO and MSEG tables and monthly FEDLOG data extracts.

rises to 93 percent and the share of the dollar value goes to 12 percent. Raising the threshold to $1,000 accounts for 94 percent of transactions and 15 percent of dollar value, and a threshold of $2,500 accounts for 97 percent of transactions and 26 percent of dollar value. As the share of the dollar value of transactions increases, Army commands would need to have more prior-year funding set aside to cover the costs of processing these transactions.

The BOP courses of action include BOP A (DoD-wide), BOP B (absorbing price changes in the AWCF for all transactions that pass through the AWCF), and BOP C (absorbing price changes in the AWCF for AMI only). To estimate the effects of absorbing price changes in the AWCF, we had to separate AMI and NAMI transactions, and also identify NAMI 71 POs, which pass through the AWCF. Table 5.3 shows the results of these calculations for FORSCOM units. BOP C accounts for only 15 percent of transactions with price increases but 73 percent of the dollar value, because AMI include most of the expensive DLRs purchased by FORSCOM units. BOP B would account for 95 percent of UMTs and 98 percent of the dollar value, because most FORSCOM units are supported by an SSA.[3] DoD-wide BOP (BOP A) would cover all UMTs (including 45 POs for NAMI), but because this course of action was recently rejected by the other services and DLA, it is most likely infeasible in the short run.

However, some of the other Army commands might not see as much benefit from BOP B and BOP C as FORSCOM. For example, ARNG and USARC units are less likely to be supported by an SSA, so they have a larger proportion of 45 POs than FORSCOM units. The results of our BOP calculations for ARNG and USARC are shown in Table 5.4. As the table indicates, BOP C would account for 8 percent of UMTs and 69 percent of the dollar value of price changes, whereas BOP B would account for only 12 percent of UMTs and 70 percent of the dollar value.

We also estimated how much the AWCF surcharge would need to increase to cover the costs of absorbing price changes in the AWCF. Over the period from October 2018 to June 2022, the total value of AMI and NAMI 71 PO price increases (net of price decreases) across all Army commands was $228.7 million, in comparison with $31,764.3 million in AWCF sales. Therefore, we estimate that the AWCF surcharge would need to increase by 0.72 percent to cover the

TABLE 5.3

U.S. Army Forces Command Transactions Affected by Born-On Pricing Alternatives

Course of Action	Number of POs with Price Increases	Percentage of POs Affected	Net Value of Price Increases	Percentage of Net Value Affected
Status quo	215,837		$110.1m	
AWCF absorbs AMI price changes (BOP C)	184,501	15	$29.4m	73
Add 71 POs for NAMI (BOP B)	11,406	95	$1.7m	98

SOURCE: Features information from GCSS-A EKPO and MSEG data tables and monthly FEDLOG data extracts.

---

[3] If BOP B was implemented in conjunction with raising the threshold for automatic processing of UMTs to $750, the two policies would account for 99.8 percent of FORSCOM transactions with price increases.

TABLE 5.4

Army National Guard and U.S. Army Reserve Command Transactions Affected by Born-On Pricing Alternatives

Course of Action	Number of POs with Price Increases	Percentage of POs Affected	Net Value of Price Increases	Percentage of Net Value Affected
Status Quo	332,755		$76.6m	
AWCF absorbs AMI price changes (BOP C)	305,931	8	$23.7m	69
Add 71 POs for NAMI (BOP B)	293,734	12	$22.8m	70

SOURCE: Features information from GCSS-A EKPO and MSEG data tables and monthly FEDLOG data extracts.

costs of BOP B. The total value of AMI price increases over the same period is $181.9 million, which would require a 0.57 percent increase in the AWCF surcharge to cover the costs of BOP C.

Conclusions on Price Changes

We found that raising the threshold for automatic processing of UMTs to $750 would cover 90 percent of UMTs. However, financial managers would still need to ensure that sufficient prior-year funding was available in the correct lines of accounting to process these transactions. As of October 2022, FORSCOM was planning to implement software that would allow for automatic processing of price changes below $750, so this course of action is feasible within existing logistics and financial management systems.

For FORSCOM, BOP B (absorbing price changes on all transactions that pass through the AWCF) would cover 95 percent of UMTs and 98 percent of the dollar value of price changes, with very little need for prior-year funding to cover the remaining UMTs. The benefits to other Army commands vary depending on their use of 45 POs, but all would pay the increase in the AWCF surcharge. BOP B (absorbing AMI price changes in the AWCF) would cover only 15 percent of FORSCOM UMTs, but 73 percent of their dollar value. Based on discussions with stakeholders, the required changes would be complex in existing logistics and financial management systems, but could be implemented more easily in EBS-C.

Policies Addressing Reliance on Credits

Among the courses of action intended to reduce reliance on credits, some stakeholders expressed concern about the SP alternative, because it eliminates financial incentives to return unserviceable DLR carcasses. To address this concern, we examined whether the credit shutoff in FY 2021 affected Army-wide returns of DLRs, in comparison with other recent years. Figure 5.3. summarizes the supply transactions that occur when a unit orders a DLR in GCSS-A. When a part reservation for a DLR is created in GCSS-A, the system asks the user whether there is an unserviceable DLR to return. When the DLR is issued from the SSA, GCSS-A then generates a return PR for an unserviceable carcass. The unit creates a return PO when it is ready to return the unserviceable DLR and then schedules a time to

FIGURE 5.3

Supply Transactions to Return Unserviceable Depot-Level Reparable Parts

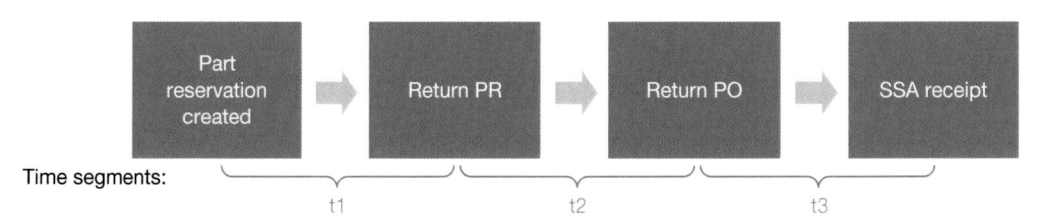

SOURCE: Adapted from unpublished RAND research by James R. Broyles, Kenneth Girardini, Candice Miller, Josh Kerrigan, and Erin Leidy.

return it to the SSA. The transaction is completed when the unserviceable DLR is received by the SSA. We label the time segments between each step as *t1, t2,* and *t3.*

The results of our data analysis are shown in Figures 5.4 through 5.6. Figure 5.4 shows the number of return PRs created from March through September 2021 (when credits were turned off, shown in the orange block) in comparison with a longer time period from January 2018 through March 2022. The return PRs are grouped by the source of supply: U.S. Army Aviation and Missile Command (AMCOM), U.S. Army Communications-Electronics Command (CECOM), and U.S. Army Tank-Automotive and Armaments Command (TACOM). Although we see a drop-off in return PRs early in the 2021 period, particularly for TACOM DLRs, the number had recovered before the end of FY 2021. In addition, these fluctuations are not unusual compared with the longer time period.

Figure 5.5 shows the percentage of return PRs converted to POs within 60 days. In other words, it indicates the share of transactions in which the unit indicated it was ready to return the DLR carcass within 60 days after the DLR was ordered ($t2 < 60$). In this case, we see some drops early in the 2021 period for AMCOM and TACOM DLRs, but again, the percentage recovers later in the period and continues the slight upward trend observed over the longer period of time.

Figure 5.6 shows the average time it took from the creation of a return PR until the unserviceable DLR carcass was received by the SSA ($t2 + t3$). We see a large increase in the average number of days to return CECOM DLRs in February to April 2021 and a smaller increase in the average time to return AMCOM DLRs. However, on average, these carcasses were still returned by the end of the FY, when credits were still turned off, and average times declined later in the FY. In addition, these patterns do not seem to be out of line in comparison with previous years.

Conclusions on Credits

We did not find any evidence that units reduced returns of DLR carcasses in response to the credit shutoff in FY 2021.[4] These results suggest that SP would be a feasible alternative to

---

[4] However, it is possible that if credits were shut off permanently, returning DLR carcasses might become a lower priority for Army units. The Army would still need to track whether the number of DLR returns is

FIGURE 5.4

Number of Return Purchase Requests Created, January 2018 to March 2022

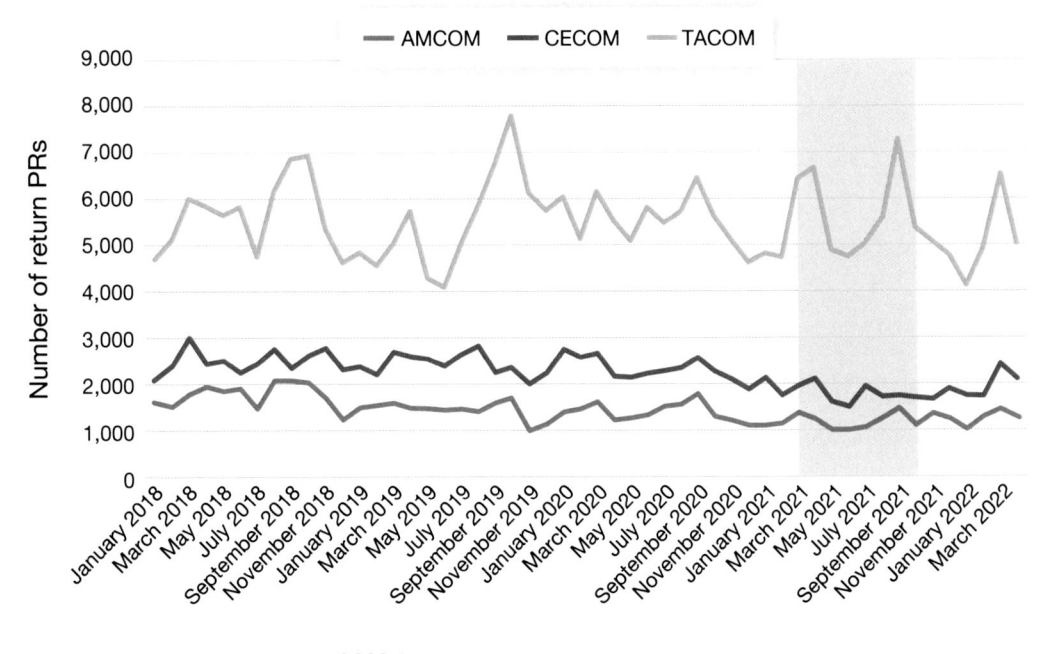

SOURCE: Features information from GCSS-A.

current price and credit policies. In addition, it could most likely be implemented in existing logistics and financial management systems with changes in the FEDLOG prices charged to Army customers. If some stakeholders still have concerns about not offering financial incentives for returning DLR carcasses, EP is also feasible. However, it would be more difficult to implement in existing logistics and financial management systems, and implementation might need to wait until EBS-C is fielded.

Evaluation of Courses of Action

In this section, we discuss our evaluation of the courses of action based on the criteria introduced in Chapter 4. We first review our evaluation of two of the most promising courses of action and then provide an overall assessment of all the courses of action.

Our evaluation of BOP B (absorbing price changes on all transactions that pass through the AWCF) is summarized in Table 5.5. For each of the evaluation criteria, we provide a rating and a brief explanation for the rating. For example, we expect BOP B to reduce the financial uncertainty associated with price changes and eliminate most of the financial management

---

consistent with the number of DLR issues.

FIGURE 5.5

Percentage of Return Purchase Requests Converted to Purchase Orders Within 60 Days

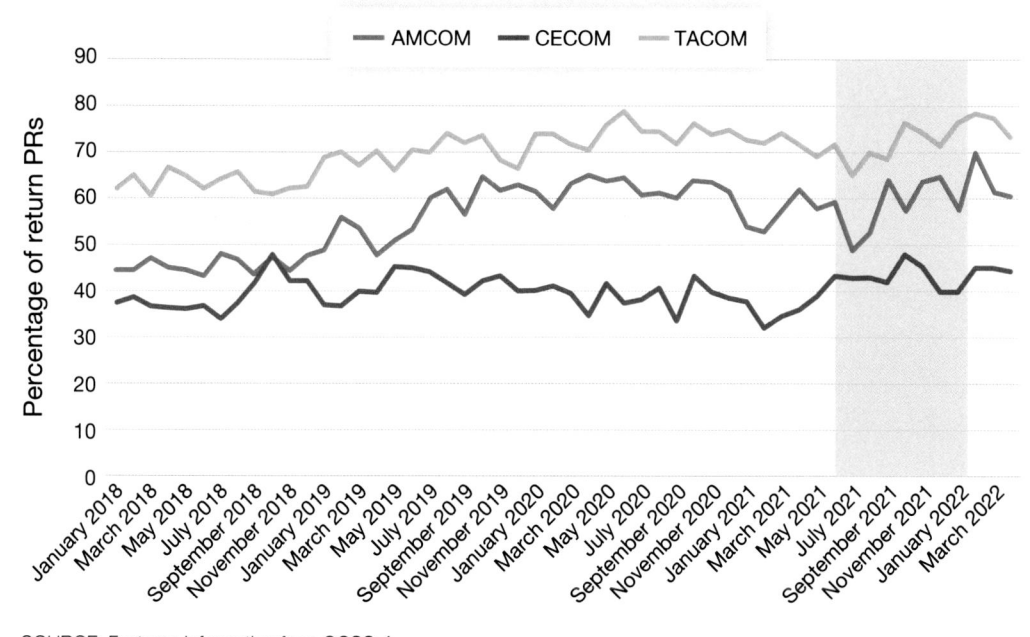

SOURCE: Features information from GCSS-A.

workload associated with processing UMTs. It would have minimal effects on DLR demands and returns, because it is not a fundamental change to existing price and credit policies. The primary drawback for BOP B is that it would be difficult to implement in existing logistics and financial management systems.

Table 5.6 summarizes our evaluation of SP, which would set a price for DLRs net of the unserviceable credit and rely on supply discipline rather than financial incentives to induce units to return unserviceable DLR carcasses. We expect SP to improve unit buying power because it results in smoother execution of units' budgets without credits coming back into their accounts. It also reduces financial management workload associated with monitoring credits. Using analysis of recent data, we estimate that removing the financial incentive would have little effect on DLR demands or carcass returns. However, the Army should continue to monitor carcass returns relative to DLR issues to ensure that carcasses are coming back for repair, because there could be financial risks to the AWCF if carcass returns are delayed or reduced and DLRs must be replaced instead of repaired to maintain supply availability. If this occurs, the AWCF might need to raise DLR prices to account for the additional costs. In addition, the lack of financial incentive for carcass returns could cause concerns on the part of AMC and OUSD(C). SP should be relatively easy to implement and likely only requires changes to the catalog price for Army customers.

FIGURE 5.6

Average Time from Return Purchase Request to Supply Support Activity Receipt

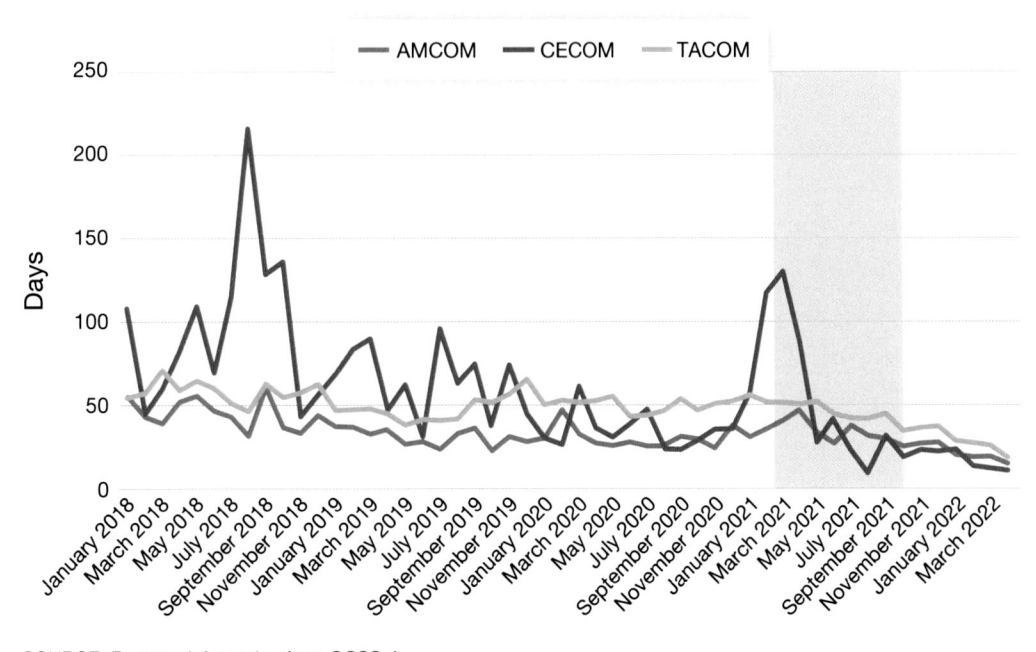

SOURCE: Features information from GCSS-A.

TABLE 5.5

Example 1: Evaluation of Born-On Pricing B

Evaluation Criteria	Rating	Explanation
Unit buying power	++	Reduces financial uncertainty of price changes
Financial management workload	++	Eliminates most UMT workload
DLR demands and returns	~	Minimal effect
Financial stability of AWCF	~	Increase surcharge (small additional risk)
Required approvals	+	Internal to the Army; would need buy-in from AMC
Changes to ERPs	X	Will require changes to ERPs
Changes to budgeting, funds distribution, and/or TRM	~	Would require slightly higher surcharge, otherwise very similar to status quo
Other services and Army reserve components	~/+	Neutral to other services; smaller benefit to Army reserve components

NOTE: Green plus signs indicate an improvement, tildes indicate minimal or no effect, and red xs indicate potential problems or implementation issues.

TABLE 5.6

Example 2: Evaluation of Single Price

Evaluation Criteria	Rating	Explanation
Unit buying power	++	Smoother execution of budgets (no credits)
Financial management workload	++	No need to monitor credits
DLR demands and returns	~	Minimal effect, based on recent experience
Financial stability of AWCF	?	Possible risk if carcasses are delayed or not returned
Required approvals	x	Internal to the Army; would need buy-in from AMC
Changes to ERPs	+	Likely only requires changes to catalog price
Changes to budgeting, funds distribution, and/or TRM	~	Minimal, but DLR prices could increase if carcass returns are lower than expected
Other services and Army reserve components	~/++	Neutral to other services; benefits Army reserve components

NOTE: Green plus signs indicate an improvement, tildes indicate minimal or no effect, and red xs indicate potential problems or implementation issues. A question mark means the effect is uncertain.

Table 5.7 summarizes our ratings of all the courses of action.[5] Among the alternatives designed to address price changes, BOP A (DoD-wide BOP) is likely infeasible in the short term, because it was recently rejected by the other services and DLA. BOP B (absorbing price changes on all transactions that pass through the AWCF) is preferred to BOP A (absorbing price changes in the AWCF for AMI only) because it provides greater benefits in terms of eliminating financial management workload and reducing the effects of price changes on units' OMA budgets. However, either of these courses of action would require changes to existing or future ERPs. Therefore, in the short term, FORSCOM could implement RT (raising the threshold for automatic processing of UMTs to $750) to gain some of the benefits of reduced financial management workload. As of October 2022, FORSCOM was already planning to make software changes to implement RT.

SP (net of unserviceable credit) and EP are the courses of action that most directly address reliance on credits. Units would pay the same price initially, but EP would impose a penalty (equal to the unserviceable credit) if the unit failed to return a DLR carcass within a fixed time period after the issue of a serviceable DLR, whereas SP would not. Like the Air Force and the Navy, the Army could continue to charge standard price and issue separate credits for non-Army customers under either SP or EP. Both policies reduce reliance on credits and fluctuations in unit budgets. The advantage of SP is that it is easier to implement in existing or future ERPs, whereas EP is likely to be more acceptable to AMC and OUSD(C) because it preserves financial incentives to return unserviceable DLR carcasses.

---

[5] More detailed explanations of the ratings for the remaining courses of action are provided in the appendix.

Table 5.7. Summary Assessment of Courses of Action

Evaluation Criteria	BOP A	BOP B	BOP C	RT	SP	EP	MPS	FI
Unit buying power	++	++	+	~	++	++	~	N/A
Financial management workload	++	++	+	+	++	+	+	+
DLR demands and returns	~	~	~	~	~	~	~	xx
Financial stability of AWCF	~	~	~	~	?	~	~	N/A
Required approvals	xx	+	+	++	x	+	+	xx
Changes to ERPs	x	x	x	+	+	x	x	??
Changes to budgeting, funds distribution, and/ or TRM	~	~	~	~	~	~	~	xx
Other services and Army reserve components	+	~/+	~/+	~	~/++	~/+	~/+	~/xx

NOTE: Green plus signs indicate an improvement, tildes indicate minimal or no effect, and red xs indicate potential problems or implementation issues. N/A = not applicable.

The remaining two courses of action address price changes and reliance on credits in very different ways. MPS would increase the visibility of shop stock to the wholesale supply system and reduce problems related to carcass matching, because issues from the SSA or shop stock could be matched with returns on a one-for-one basis. AMC is already exploring the feasibility of implementing MPS for some units in existing ERP systems, and it could be implemented more broadly as part of EBS-C. In addition, the improved inventory visibility and control associated with MPS may be a prerequisite for FI or a policy similar to the Air Force's CPFH pricing, if units no longer face the budget constraints of buying individual DLRs with OMA funding.

There would be many uncertainties involved if the Army wanted to move toward FI or a CPFH model. Additional data analysis and pilot testing would most likely be needed to determine whether either of these alternatives is feasible. For example, how well do helicopter flying hours or tank miles predict DLR purchases? What additional analysis and modeling would be needed to set a CPFH or a cost per vehicle mile by weapon system that would allow the AWCF to break even? It would also be difficult to forecast changes in customer behavior if units no longer had to pay for individual DLRs. The Army might need alternative processes to allocate DLRs in short supply to ensure that they went to the highest-priority units.

CHAPTER 6

Findings and Recommendations

In this chapter, we summarize our findings and recommendations regarding changes to Army price and credit policies for spare parts.

Findings

The AWCF experienced cash flow problems beginning in FY 2020 because of a reduction in sales caused by the COVID-19 pandemic that occurred at the same time the AWCF had increased purchases to address supply availability problems. As a result, the AWCF turned off credits to customers in Army units for most DLRs from March through September 2021. We estimated that Army-wide customers faced a reduction of approximately $860 million in OMA funding because of the credit shutoff. The AWCF also received nearly $2 billion in cash infusions, increased its surcharge, and reduced purchases from FY 2020 to FY 2022, and the cash balance was restored by the end of FY 2022.

Customers in Army units continued to return DLRs to the supply system during the period that credits were turned off, even though they no longer had a financial incentive to do so. Therefore, we concluded that an SP policy, set at the standard price minus the credit, could be feasible to reduce reliance on credits. The EP policy would function in a similar way, because customers pay the standard price minus the credit when they order a DLR, but the financial incentive is preserved, because they are charged a delta bill equal to the credit if they do not return a DLR carcass within a fixed time period, such as 60 days.[1]

Moving the point of sale to the job order would increase visibility of DLRs currently held in Plant 2000 (including shop stock) and reduce incentives to hold DLR carcasses until there is a matching issue, because issues and returns of DLRs would occur on a one-for-one basis. However, this policy by itself would not reduce reliance on credits. Under a policy of FI of DLRs, units would no longer be required to pay for DLRs or receive credits for returning carcasses. However, the potential effects on DLR demands are unpredictable, and the Army might need an alternative mechanism to ensure that scarce DLRs are sent to the highest-priority customers.

---

[1] Policies that reduce reliance on credits would not prevent the AWCF from having cash flow problems in the future. Other policy levers would be needed to address a cash shortage.

We also examined price and credit policies that would reduce the effects of price changes on Army units. Although the DoD FMR states that prices are supposed to be stabilized during the FY, there are many exceptions that result in mid-year price changes. In addition, nearly all prices on outstanding orders are revalued at the beginning of each FY, and customers must have prior-year funds available to pay for any price increases. The resulting unmatched transactions create financial management workload, primarily at the division, corps, and Army command levels. In addition, FORSCOM and its units try to fully execute their OMA funding prior to the end of the FY, to reduce the risk that the funding will never be spent, so funds to cover UMTs are often scarce.

We found that raising the threshold for automatic processing of UMTs from $250 to $750 would cover 90 percent of transactions with a cost of about $13 million over three years. Financial managers would still need to ensure that prior-year funding is available in the correct line of accounting, but this policy would be relatively easy to implement in existing ERP systems. BOP policies that set the price at the time a spare part is ordered would be more effective at reducing the effects of price changes. DoD-wide BOP (BOP A) requires agreement from the other services and DLA, and it was recently rejected by a DoD-wide integrated product team, so it is unlikely to be implemented in the near future. However, the Army would not need consent from the other services to absorb some price changes in the AWCF. BOP for AMI only (BOP C) would cover about 15 percent of FORSCOM's UMTs and 73 percent of the dollar value, and the effects would be similar for other commands. The AWCF could also absorb price changes on NAMI if the transactions pass through an SSA (BOP B). This broader form of BOP would cover 95 percent of FORSCOM's UMTs and 98 percent of the dollar value. However, it would be less effective for some Army commands whose units are less likely to be supported by SSAs, such as the ARNG and USARC. We estimated that BOP C would require an increase in the AWCF surcharge of 0.57 percent, whereas BOP B requires an increase of 0.72 percent.

Recommendations

To address reliance on credits, the most promising courses of action are SP or EP. SP could most likely be implemented in existing ERP systems by changing the catalog price paid by Army customers. It would eliminate financial incentives to return DLR carcasses, but there is evidence that customers continued to return carcasses during the period when credits were shut off in FY 2021. However, there might still be concerns that returning carcasses could become a lower priority for units, and there may eventually be fewer carcasses available for depot-level repair programs. Therefore, the Army would need to continue to monitor carcass returns to ensure that they remain in balance with DLR issues. EP would preserve the financial incentive to return DLR carcasses, but it would be more difficult to implement in ERP systems than SP and may need to wait until EBS-C is fielded.

To address price changes, FORSCOM and other Army commands can raise the threshold for automatic processing of UMTs to price changes up to $750 in the short term. This policy would still require financial managers to ensure that prior-year funding is available in the correct line of accounting, but it can be implemented in existing ERP systems. In the longer term, we recommend that the Army implement BOP B, absorbing price changes for AMI and NAMI transactions that pass through the AWCF. This policy would eliminate almost all UMTs for FORSCOM and would require a relatively small increase in the AWCF surcharge. However, it would be complex to implement in ERP systems and may need to wait until EBS-C is fielded.

Evaluation of Remaining Courses of Action

In this appendix, we provide additional information about our evaluation of the courses of action that were not discussed in detail in the main text.

Our evaluation of BOP A, DoD-wide BOP, is summarized in Table A.1. Prices would be set at the time a spare part is ordered, so all UMTs and financial uncertainty associated with price changes would be eliminated. Small increases in the AWCF and other WCF surcharges would be needed to absorb the effects of changes in the purchase prices or repair costs of the items they manage. However, this course of action was recently rejected by the other services and DLA, and would require changes to ERP systems.

Table A.2 summarizes our evaluation of BOP C, which would absorb price changes on AMI in the AWCF. It is less effective at reducing financial uncertainty and eliminating UMTs than BOP A and BOP B, because it covers a smaller share of UMTs, but it would cover about 70 percent of the dollar value of price changes, because DLRs are typically more expensive than consumables. Because this course of action does not affect the other services, approvals would be internal to the Army, but it would require changes to ERP systems and a small increase in the AWCF surcharge.

TABLE A.1

Evaluation of Born-On Pricing A

Evaluation Criteria	Rating	Explanation
Unit buying power	++	Eliminates financial uncertainty of price changes
Financial management workload	++	Eliminates UMT workload
DLR demands and returns	~	Minimal effect
Financial stability of AWCF	~	Increase surcharge (small additional risk)
Required approvals	xx	Requires approval of other services and DLA
Changes to ERPs	x	Will require changes to ERPs
Changes to budgeting, funds distribution, and/or TRM	~	Would require slightly higher surcharge, otherwise very similar to status quo
Other services and Army reserve components	+	Would benefit other services and Army reserve components

NOTE: Green plus signs indicate an improvement, tildes indicate minimal or no effect, and red xs indicate potential problems or implementation issues.

TABLE A.2

Evaluation of Born-On Pricing C

Evaluation Criteria	Rating	Explanation
Unit buying power	+	Reduces financial uncertainty of price changes for AMI only
Financial management workload	+	Eliminates UMT workload for AMI
DLR demands and returns	~	Minimal effect
Financial stability of AWCF	~	Increase surcharge (small additional risk)
Required approvals	+	Internal to the Army; would need buy-in from AMC
Changes to ERPs	X	Will require changes to ERPs
Changes to budgeting, funds distribution, and/or TRM	~	Would require slightly higher surcharge, otherwise very similar to status quo
Other services and Army reserve components	~/+	Neutral to other services; benefits Army reserve components

NOTE: Green plus signs indicate an improvement, tildes indicate minimal or no effect, and red xs indicate potential problems or implementation issues.

Table A.3 summarizes our evaluation of RT. This course of action reduces some of the workload associated with UMTs, but price changes would still have to be funded by customers or higher commands, and financial managers would need to ensure that prior-year funding is available in the correct line of accounting. The decision to implement this course of action would be internal to FORSCOM (or other commands that chose to implement it), and it is relatively easy to implement in existing ERP systems.

Table A.4 summarizes our evaluation of an EP policy. Customers would be charged the standard price minus the credit when they buy a DLR, but would be charged a delta bill if they fail to return an unserviceable DLR carcass within a fixed period of time, such as 60 days. This policy would result in a smoother execution of OMA budgets, because no credits are flowing back into unit accounts, but it still requires units to match carcass returns to issues to avoid delta bills. This policy might be more acceptable to OUSD(C) and AMC, because it preserves financial incentives to return DLR carcasses. However, it would require changes to ERP systems.

Our evaluation of MPS is summarized in Table A.5. This course of action would not affect current price and credit policy, but it would enable one-for-one matching of DLR issues and carcass returns, because OMA-funded inventories held in Plant 2000 (including shop stock) would be incorporated into the AWCF and be visible to wholesale supply managers. Thus, units would have less reason to hold onto DLR carcasses if they did not have a matching issue. The decision to implement this course of action would be internal to the Army, and AMC is exploring the feasibility of implementing it for some units in existing ERP systems. Moving the point of sale could be implemented more broadly in EBS-C.

Table A.6 summarizes our evaluation of FI. Under this course of action, Army units would no longer need budgets for DLRs and would not need to monitor price changes or credits for

TABLE A.3

Evaluation of Raising the Threshold for Automatic Processing of Unmatched Transactions

Evaluation Criteria	Rating	Explanation
Unit buying power	~	Price changes still affect units or higher echelons
Financial management workload	+	Reduces some UMT workload
DLR demands and returns	~	No effect
Financial stability of AWCF	~	No effect
Required approvals	++	Internal to FORSCOM
Changes to ERPs	+	Requires minimal software changes
Changes to budgeting, funds distribution, and/or TRM	~	No effect
Other services and Army reserve components	~	No effect on other services or Army reserve components

NOTE: Green plus signs indicate an improvement and tildes indicate minimal or no effect.

TABLE A.4

Evaluation of Exchange Pricing

Evaluation Criteria	Rating	Explanation
Unit buying power	++	Smoother execution of budgets (no credits)
Financial management workload	+	Need to monitor carcass returns to avoid delta bills
DLR demands and returns	~	No effect, financial incentives same as status quo
Financial stability of AWCF	~	No effect
Required approvals	+	Internal to the Army; would need buy-in from AMC
Changes to ERPs	x	Will require changes to ERPs
Changes to budgeting, funds distribution, and/or TRM	~	No effect
Other services and Army reserve components	~/+	Neutral to other services; benefits Army reserve components

NOTE: Green plus signs indicate an improvement, tildes indicate minimal or no effect, and red xs indicate potential problems or implementation issues.

DLRs. However, they would still need OMA budgets for items managed by the other services, DLA, and GSA and would still be affected by price changes on these items. There could potentially be large increases in demands for DLRs if units no longer had budget constraints on ordering DLRs, and the Army might need alternative mechanisms to ensure that DLRs in short supply are sent to the highest-priority customers. Broad changes to budgeting, funds distribution, and ERP systems would be needed if funding for DLRs went to AMC instead of

TABLE A.5

Evaluation of Moving the Point of Sale to the Job Order

Evaluation Criteria	Rating	Explanation
Unit buying power	~	Does not affect current price and credit policy
Financial management workload	+	Enables one-for-one matching of DLR issues and carcass returns
DLR demands and returns	~	Minimal effect, could reduce some current delays in DLR returns
Financial stability of AWCF	~	Minimal effect, possible benefit from increased inventory visibility
Required approvals	+	Internal to the Army
Changes to ERPs	x	Will require changes to ERPs
Changes to budgeting, funds distribution, and/or TRM	~	Minimal effect, but reduces financial buffer of stocks owned in OMA
Other services and Army reserve components	~/+	Neutral to other services; benefits Army reserve components

NOTE: Green plus signs indicate an improvement, tildes indicate minimal or no effect, and red *xs* indicate potential problems or implementation issues.

TABLE A.6

Evaluation of Free Issue

Evaluation Criteria	Rating	Explanation
Unit buying power	N/A	Units would no longer need OMA budgets for DLRs
Financial management workload	+	No need to monitor price changes or credits
DLR demands and returns	xx	Uncertain effects if units no longer have budget constraints on purchasing DLRs
Financial stability of AWCF	N/A	AMC would be directly funded to purchase and repair DLRs
Required approvals	xx	Would require approval from OSD
Changes to ERPs	??	Unclear how ERPs would need to be changed if issues and returns of DLRs did not require financial transactions for Army customers
Changes to budgeting, funds distribution, and/or TRM	xx	Funding for DLRs would go to AMC instead of units; could become more difficult to forecast DLR demands
Other services and Army reserve components	~/xx	Neutral to other services if standard price and credit still apply to them/Army reserve components could have lower priority for DLRs in short supply

NOTE: Green plus signs indicate an improvement, tildes indicate minimal or no effect, and red *xs* indicate potential problems or implementation issues.

units and financial transactions were not required for AMI but were still needed for NAMI. In addition, this course of action would most likely require approval from OUSD(C).

Abbreviations

ABCT	armored brigade combat teams
AFWCF	Air Force Working Capital Fund
AMC	U.S. Army Materiel Command
AMCOM	U.S. Army Aviation and Missile Command
AMI	Army-managed items
ARNG	Army National Guard
AWCF	Army Working Capital Fund
BOP	born-on pricing
CAB	combat aviation brigade
CAM	centralized asset management
CECOM	U.S. Army Communications-Electronics Command
COVID-19	coronavirus disease 2019
CPFH	cost per flying hour
DASD(L)	Deputy Assistant Secretary of Defense for Logistics
DLA	Defense Logistics Agency
DLR	depot-level reparable part
DoD	U.S. Department of Defense
EBS-C	Enterprise Business Systems—Convergence
EP	exchange price
ERP	enterprise resource planning
FEDLOG	Federal Logistics Data
FI	free issue
FMR	Financial Management Regulation
FORSCOM	U.S. Army Forces Command
FY	fiscal year
GCSS-A	Global Combat Support System-Army
GFEBS	General Fund Enterprise Business Systems
GSA	General Services Administration
IBCT	infantry brigade combat team
MPS	moving the point of sale to the job order
NAMI	non–Army-managed items
NIIN	National Item Identification Number
O&M	operations and maintenance
OMA	Operations and Maintenance, Army

OPTEMPO	operations tempo
OSD	Office of the Secretary of Defense
OUSD(C)	Office of the Undersecretary of Defense (Comptroller)
PO	purchase order
PR	purchase request
RT	raising the threshold for automatic processing of unmatched transactions
SAF/FMB	Deputy Assistant Secretary of the Air Force for Budget
SBCT	Stryker brigade combat team
SP	single price
SSA	Supply Support Activity
TACOM	U.S. Army Tank-Automotive and Armaments Command
TRM	Training Resource Model
UMT	unmatched transaction
USARC	U.S. Army Reserve Command
WCF	working capital fund

References

Baldenius, Tim, and Stefan Reichelstein, "External and Internal Pricing in Multidivisional Firms," *Journal of Accounting Research,* Vol. 44, No. 1, March 2006.

Baldwin, Laura H., and Glenn A. Gotz, *Transfer Pricing for Air Force Depot-Level Reparables,* RAND Corporation, MR-808-AF, 1998. As of November 10, 2021: http://www.rand.org/pubs/monograph_reports/MR808.html

Bozin, Stanley D., *Implementation of Improved Management Control of Aviation Depot Level Repairable Funds,* thesis, Naval Postgraduate School, December 1986.

Brauner, Marygail K., Ellen M. Pint, John R. Bondanella, Daniel A. Relles, and Paul Steinberg, *Dollars and Sense: A Process Improvement Approach to Logistics Financial Management,* MR-1131-A, RAND Corporation, 2000. As of November 10, 2021: http://www.rand.org/pubs/monograph_reports/MR1131.html

Byrnes, Patricia E., "Defense Business Operations Fund: Description and Implementation Issues," *Public Budgeting & Finance,* Vol. 13, No. 4, December 1993.

Davis, Carl, "EP CBA Final Results," PowerPoint briefing, March 2013, Not available to the general public.

Department of the Army, *Army Working Capital Fund, Fiscal Year (FY) 2013 President's Budget,* February 2012.

Department of the Army, *Army Working Capital Fund Fiscal Year (FY) 2014 Budget Estimates,* April 2013.

Department of the Army, *Army Working Capital Fund Fiscal Year (FY) 2015 Budget Estimates,* March 2014.

Department of the Army, *Army Working Capital Fund Fiscal Year (FY) 2016 Budget Estimates,* February 2015.

Department of the Army, *Army Working Capital Fund Fiscal Year (FY) 2017 Budget Estimates,* February 2016.

Department of the Army, *Army Working Capital Fund Fiscal Year (FY) 2018 Budget Estimates,* May 2017.

Department of the Army, *Army Working Capital Fund Fiscal Year (FY) 2019 Budget Estimates,* February 2018.

Department of the Army, *Army Working Capital Fund Fiscal Year (FY) 2020 Budget Estimates,* March 2019.

Department of the Army, *Army Working Capital Fund Fiscal Year (FY) 2021 Budget Estimates,* February 2020.

Department of the Army, *Army Working Capital Fund Fiscal Year (FY) 2022 Budget Estimates,* May 2021.

Department of the Army, *Army Working Capital Fund (FY) Fiscal Year 2023 Budget Estimates,* April 2022.

Department of the Navy, *Depot Level Repairable Item Management,* OPNAV Instruction 4400.9D, September 18, 2017.

DoD FMR—*See* Under Secretary of Defense (Comptroller).

Hirschleifer, Jack, "On the Economics of Transfer Pricing," *Journal of Business,* Vol. 29, No. 3, 1956.

Jordan, Leland G., "Defense Business Operations Fund (DBOF): Problems and Promise," *Public Budgeting & Finance,* Vol. 15, No. 4, December 1995.

Lang, Thomas E., and Joseph N. Kader, *Born-On Pricing/Unmatched Transactions Analysis,* LMI, April 2020.

Light, Thomas, Michael Boito, Tim Conley, Larry Klapper, and John Wallace, *Understanding Changes in U.S. Air Force Aircraft Depot-Level Reparable Costs over Time,* RAND Corporation, 2018, Not available to the general public.

Peltz, Eric, Marygail K. Brauner, Edward G. Keating, Evan Saltzman, Daniel Tremblay, and Patricia Boren, *DoD Depot-Level Reparable Supply Chain Management: Process Effectiveness and Opportunities for Improvement,* RAND Corporation, RR-398-OSD, 2014. As of February 28, 2023:
https://www.rand.org/pubs/research_reports/RR398.html

Pint, Ellen M., Marygail K. Brauner, John R. Bondanella, Daniel A. Relles, and Paul Steinberg, *Right Price, Fair Credit: Criteria to Improve Financial Incentives for Army Logistics Decisions,* RAND Corporation, MR-1150-A, 2002. As of March 1, 2023:
https://www.rand.org/pubs/monograph_reports/MR1150.html

Under Secretary of Defense (Comptroller), *Department of Defense Financial Management Regulation:* Vol. 2B, *Budget Formulation and Presentation (Chapters 4–19),* DoD 7000.14-R, last updated September 2022a.

Under Secretary of Defense (Comptroller), *Department of Defense Financial Management Regulation:* Vol. 11B, *Reimbursable Operations Policy—Working Capital Funds (WCF),* DoD 7000.14-R, last updated May 2022b.

U.S. Air Force, "United States Air Force Working Capital Fund (Appropriation: 4930): Fiscal Year (FY) 2021 Budget Estimates," February 2020.

U.S. Air Force, "United States Air Force Working Capital Fund (Appropriation: 4930): Fiscal Year (FY) 2022 Budget Estimates," May 2021.

U.S. Air Force, "United States Air Force Working Capital Fund (Appropriation: 4930): Fiscal Year (FY) 2023 Budget Estimates," April 2022.

U.S. Army War College, *How the Army Runs: A Senior Leader Reference Handbook, 2021–2022,* December 27, 2021.

U.S. Government Accountability Office, *Military Training: Actions Needed to More Fully Develop the Army's Strategy for Training Modular Brigades and Address Implementation Challenges,* GAO-07-936, August 6, 2007.